1.16   Schuler, Stanley
           The Fl   nd ceiling
       ok.
                              27

THE FLOOR AND CEILING BOOK

Other building books by Stanley Schuler

THE WALL BOOK
HOW TO FIX ALMOST EVERYTHING (Revised Edition)
THE COMPLETE BOOK OF CLOSETS AND STORAGE
THE HOMEOWNER'S MINIMUM-MAINTENANCE MANUAL

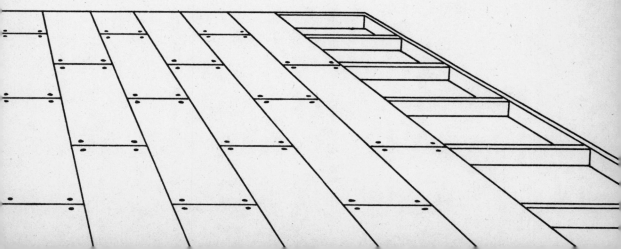

THE FLOOR AND CEILING BOOK

Everything you need to know about building, repairing, maintaining, soundproofing, finishing and decorating the floors and ceilings in your home

by STANLEY SCHULER

Illustrated by Marilyn Grastorf

M. EVANS AND COMPANY, INC.
New York, N.Y. 10017

M. Evans and Company titles are distributed in
the United States by the J. B. Lippincott Company,
East Washington Square, Philadelphia, Pa. 19105;
and in Canada by McClelland & Stewart Ltd.,
25 Hollinger Road, Toronto M4B 3G2, Ontario

LIBRARY OF CONGRESS CATALOGING IN PUBLICATION DATA

Schuler, Stanley.
    The floor and ceiling book.

    Includes index.
    1. Floors.   2. Ceilings.   I. Title
TH2521.S4   690'.1'6   75-37966
ISBN 0-87131-201-8

Design by Joel Schick

Manufactured in the United States of America

9 8 7 6 5 4 3 2 1

CONTENTS

THE FLOOR AND CEILING BOOK

1/ GETTING STARTED ON FLOOR AND CEILING WORK

Demanding is the best word I can think of to describe the floors and ceilings in a house. They never stop demanding attention.

Admittedly, this is truer of floors than of ceilings, for the simple reason that they receive much, much harder wear. Even so, if you were to add up the hours of work you devote to keeping your house neat, clean, and in good repair, you would be amazed how much more demanding ceilings are than many other parts of the house.

Floors, of course, are in a league of their own.

I shall not soon forget the depressing condition of the floors and ceilings in our present house when my wife and I bought it seven years ago. The house had been unoccupied for two or three years and hadn't received much care from its elderly owners in the preceding decade; so it wasn't surprising that many things had gone to rack if not to complete ruin. But the condition of the floors and ceilings hit us right in the face.

In *The Wall Book*, the companion volume that preceded this one, I wrote that the interior walls were the first things in the house that we tackled. And that is true. But it wasn't because they were in worse condition than anything else. They weren't. It was because the wall surfaces in the house (as in other houses) were so much larger than other surfaces—so much right in front of your eyes—that we simply had to fix them up at once. But the floors and ceilings followed right behind.

There wasn't a single floor that could be called acceptable. All were filthy with ground-in dirt, scarred, scratched, and stained. Several had big, deep burns caused by embers that had leaped from the fireplaces. In the dining room, an enormous rusty grille (once the outlet for a floor furnace) let in damp, musty air from a crawl space. In the living room several of the once-beautiful random-width pegged-oak floorboards had warped upward along the edges to form ridges about ¾ inch high. One of the halls was covered only with painted but rough subfloorboards. A bedroom also had nothing but subflooring—in this case unpainted. The kitchen floor evidently had been so dreadful that the owners, frantic to sell, had torn it out completely and replaced it with plywood—

1

not the smooth-surfaced, plugged and knot-free plywood that is normally used as underlayment for resilient flooring, but the rough stuff that is used for sheathing. The linoleum in the bathrooms was torn, brittle, and rotten. And our bedroom floor had not only sagged but also creaked and groaned under the lightest weight.

The ceilings were little better. The plaster hung in chunks from several. Others that were plastered were fissured with cracks which, in several cases, the owners had attempted to conceal with new sanded paint that was already beginning to flake off. The gypsum board between the ceiling beams in the dining room had been soaked so often by leaks in the roof deck above that they sagged disconsolately. In the laundry, water from a dripping pipe had eaten a hole right through one of the gypsum board panels.

The house was awful. But we had fallen in love with it at first sight; and though we shuddered at the work that had to be done, we were not dismayed. Time and a lot of work have confirmed our faith. But we shall never reach the point where we can sit back, laurels in hand, and tell ourselves, "Well, that's that for the floors and ceilings."

As I said at the outset, floors and ceilings never stop demanding attention. Dirt, wear, moisture, mildew, and sunlight take their toll every day. Houses settle, wood warps, materials expand and contract, decorating schemes go out of style—something is forever happening that makes homeowners decide that an old floor or ceiling needs work.

And every now and then—usually when homeowners run short of living space—they even go so far as to build a totally new floor or ceiling.

When this time comes, two questions inevitably enter the mind: What will we do? Who's going to do it?

What will we do?

In most cases, the answer is obvious: You make whatever repairs are indicated, refinish the floor or ceiling with the same or similar material, or replace the resilient floor or carpet with new material. In some cases, however, your course of action is wide open. Here's where the unwary often get into trouble, because there's a lot more to selecting a floor—and even a ceiling—than merely deciding, "I like the looks of this and don't like the looks of that."

If you're going to wind up with the desired result—whatever that may be—you must give thought to a number of points:

Grade level of subfloor.

The position of a floor in relation to the ground is an important consideration, because moisture in the ground or flowing across it is one of the worst enemies that floors have. It damages or destroys some types of flooring, prevents other types that are installed with adhesive from sticking tightly to the base underneath.

Happily, the only subfloors with which problems arise are made of poured concrete laid directly on the ground (with or without a thin layer of gravel in between). Of these, the more troublesome are below-grade subfloors which, because of their position below ground level, are especially vulnerable to flooding as well as to the fluctuating water table and hydrostatic pressure. Since on-grade subfloors are at ground level, moisture problems are likely to be somewhat less acute; nevertheless, finish flooring materials that are unsuitable for use at below-grade level are almost always equally unsuitable for use on grade.

By contrast, suspended subfloors are free of moisture problems because they have an air space underneath as well as above. So whether they are built of concrete (very rare in homes) or wood, you can cover them with any type of finish flooring.

Durability.

The ability to stand up to all the many forces to which building materials are subjected is something we look for—especially in these days of poor product quality—in all parts of a house. To be sure, it is not of great importance when you're selecting a ceiling material because the only things that can destroy a ceiling are water and the racking and settling of the house—and these are as dangerous to one material as to another. But floors are another matter: Few other parts of a house are as exposed to so much wear.

This is not the place to discuss the relative durability of the many materials used in floors. That comes in later chapters discussing the use, installation, and maintenance of specific materials. But it must be noted that there are definite differences in flooring durability, and you don't want to make the mistake of using something in the wrong place.

I knew of a man who put down cork tile in his front hall because he had a fetish about quiet houses and thought that halls outranked all other areas except the kitchen in the noisemaking department. But it wasn't long before he discovered that he'd made a poor choice. The wet and gritty shoes of the hordes passing through the hall made mincemeat of the cork in short order.

This, of course, is just one example, but it's not isolated. Homeowners the country over persist in using the wrong flooring materials in heavily trafficked areas—largely because it reduces the initial cost—and then use better-than-necessary materials in bedrooms, where the wear is minimal.

Resistance to stains.

This is of particular importance in kitchens and bathrooms, where so many different staining agents are dropped on floors, but it should not be overlooked in the front hall or family room. Nothing disfigures a floor quite as much as a large stain that defies efforts to remove it.

True, all flooring materials can be stained by something; but some have more all-around resistance than others. Vinyl, for instance, is superior to linoleum; quarry tile is better than flagstone; and almost anything is way ahead of carpet.

Ease of maintenance.

Between the two standard ceiling materials—plaster and gypsum board—there is little choice. If properly installed, one is as good as the other; if improperly installed, one is as bad as the other. Admittedly, if a house settles, plaster may crack; and once a large structural crack develops in a ceiling, it takes a long time to bring it under control. But what most people don't realize is that the warping of joists—which is about as common as settlement of an entire house—can make an equally bad mess out of gypsum board.

Ease of maintenance, however, is an urgent consideration in the selection of a floor. I was made most aware of that about twelve years ago when I was peripherally engaged in a battle between the carpet and resilient flooring industries. At that time, the huge market for floors in schools, hospitals, and commercial buildings was pretty well controlled by the resilient manufacturers. Then, thanks to the development of synthetic fibers, the carpet people took aim at the market and began to make inroads. Not content with the indisputable argument that carpet contributes to the quietness of classrooms, corridors, and so on, they began to make claims in every direction; and one of the most vociferous was that carpet is easier to maintain than any of the resilient materials. This is where I was brought into the picture. My assignment was to find out which of the antagonists was telling the truth. But short of·initiating a long-term laboratory study, it was an impossible undertaking, because carpet hadn't been in use in public buildings long enough for anyone to make a valid comparison. Every maintenance man I talked with had ideas, but nothing much to back them up. As far as I know, the argument has never been settled conclusively.

Despite this ending to the battle, the fact remains that the different flooring materials used in homes vary in the amount of maintenance they require. How much depends not only on the materials themselves but also on the rooms in which they are used, the number of people in the house, the way the people live, the way they are accustomed to taking care of their floors, their attitude toward what is acceptable and objectionable in the appearance of a floor, and so on. All these are matters that you should weigh carefully.

Quietness.

Here is a problem that merits more thought than many homeowners give it. By and large, American homes are a great deal noisier than they should be for real comfort. I don't know whether people themselves are noisier than they used to be (though children rarely any longer obey the adage that they should be seen but not heard). I'm not sure whether we have become more hardened to noise as a result of constant exposure to roaring traffic, airplanes, and factories. But I do know that, because of thinner walls and more open floor plans, houses have lost the ability to limit the transfer of sound from one room to another. And for no very clear reason (except possibly

that people really are noisier than they used to be), the noise level within rooms has risen.

The floors and ceilings you install can do much to correct the problem if you deem correction necessary. But since the subject is a bit technical and has ramifications, I'll leave further discussion to Chapter 3.

Resilience.

Resilience in flooring (you can ignore ceilings) contributes not only to the quietness of a house but also to the comfort of the people walking through the house. This is obvious. Everyone would rather walk on a carpeted floor than a bare floor because it is less jarring to the eardrums, bones, and muscles. However, since the only room in which we do very much walking and standing is the kitchen, we generally don't worry about the comfortableness of flooring except in the kitchen.

But there is more to the subject of flooring resilience than this. You should also consider the way that flooring materials react to the kinds of loads that are placed on them.

One kind of load is dynamic, or live. It occurs when a foot or a fallen object touches a floor. The weight or pressure on the floor is momentary, but it is capable of doing considerable damage at the point of impact. When spike heels were in vogue, the damage they did to floors was almost unbelievable. Concrete and tile were disintegrated; wood and linoleum dented. Even carpet and vinyl, which are able to spring back into normal shape after an impact, showed the effects.

Happily, now that spike heels are out of style, the damage caused to floors by dynamic loads is done only by heels with protruding nails and by heavy weights that are dropped. On the other hand, static loads are as much of a problem today as they have been throughout history.

A static load is a dead weight—the weight

of a chair, bed, or piano. It has little or no effect on hard-surfaced flooring but usually leaves a permanent mark on soft-surfaced flooring. The intensity of the mark depends on how long the weight is left in one place and on the resilience of the flooring materials. Asphalt tile, for example, is slower to dent than vinyl flooring because it is harder; but once dented, it is scarred forever, whereas the more resilient vinyl may regain its normal shape.

Light reflectance.

Good lighting, which allows you to walk around your home safely and see what you are doing at all times, is the product not only of efficient lamps and fixtures but also of ceilings, floors, walls, and furnishings that reflect rather than absorb the light.

For maximum reflectance, the ideal color is, of course, pure white because it reflects almost 90 percent of the light that strikes it. Black, purple, dark red, and brown are at the opposite end of the scale.

Because of its strategic overhead position, the reflectance value of the ceiling should range between 60 and 90 percent. The higher the value, the better—especially in kitchens, bathrooms, and other areas where you work. Floors may have a reflectance value of as low as 15 percent, but because your eyes are directed downward during the performance of many visual activities, values of between 25 and 35 percent are preferred.

Here are the approximate reflectance values of a number of colors used in homes:

Off white	82%
Medium gray	50%
Sky blue	49%
Spring leaf green	45%
Pale yellow	78%
Lemon	64%
Corn yellow	54%
Chocolate	15%
Saffron	35%
Very pale pink	70%
Pink	60%
Mandarin red	21%
Walnut	10%
Driftwood	69%

Reflectance is not solely influenced by color, however. Texture and finish also play parts. Rough surfaces, for instance, don't reflect as much of the light from electric fixtures as smooth surfaces. Similarly, glossy surfaces are less desirable than matte or dull surfaces because they create distracting and uncomfortable shiny reflections without, as a rule, bouncing more light back into a room.

Cost.

Cost is an inescapable consideration that bears heavily on the ultimate decision about which ceiling or flooring material to use. In view of the long life of a ceiling or floor, however, it is not one that should be allowed to overweigh other considerations unduly. In flooring materials, at least, high cost is usually a reflection of durability; and if, say, a vinyl floor costs twice as much as vinyl-asbestos but lasts half again as long, you're ahead of the game to buy it. This is particularly true of carpets. On the other hand, if you're putting in a ceiling, the lower cost of gypsum board is a very good reason for choosing it over plaster, because there is little if any difference in durability.

Appearance.

Although I put this last, the appearance of flooring—and in some cases, of ceiling materials—is usually the first thing anyone thinks about.

I doubt that many people permit their subjective feelings about the appearance of a flooring material to outweigh their better judgment. We buy the material that is going to look right in the room in which it's used even though it may not be the one we think is most beautiful. Although we may not consciously ask ourselves, "Is the color good? Is the texture right? Is the pattern suitable?"—all these questions are encompassed in our thinking.

But there are two points that are sometimes ignored.

First, unless there are very compelling reasons to the contrary, it is usually a poor idea to use a lot of different flooring materials on the same level of a house. This doesn't mean that the floors on the first story must match those on the second story (although it's better if they do). And I most certainly do not mean that the floors in the kitchen and bathrooms should be the same as those in the living areas and bedrooms. In fact, if you want a brick floor in a front hall while you use wood floors in the adjacent rooms, that's quite permissible. Or if you want slate in a family room that opens off a living room with a wood floor, that's OK, too.

But when you wind up with a motley collection of floors—especially if they are all made of the same basic material—the effect is disturbing. In two of the old houses I have owned, for example, all the floors (except in the kitchen and bathrooms) were wood; but they had no other common denominator. I had oak flooring here; spruce flooring there; old wide-board pine flooring elsewhere—and they just didn't go well together.

To look its best a house needs something that ties the various parts and rooms together. And because the walls are usually treated differently from room to room, this role is assumed by the floors—which are the next largest and most prominent surfaces in the house—and by the ceilings—which are the third most prominent surfaces.

The second—and less important—point that is often ignored when we consider flooring from an appearance standpoint is the effect that sunlight has on colors.

All flooring materials other than ceramic tile lose color to a certain extent when they are used in rooms with a wide-open southern exposure. Contrary to some belief, the windows don't have to be large. In our previous house we had windows of ordinary size on the south side, but because we had no trees to shade them, the sunlight pouring through was devastating. Within a year we could see a marked change in the color of the oak flooring and acrylic carpet.

The same sort of thing also happens to floors in rooms with eastern and western exposures. In these cases, however, trouble usually occurs only when the windows are enormous, and even then the color change is gradual.

Obviously, the best way to prevent fading is to shade southern windows. But you can also minimize the problem by selecting floor colors with above-average resistance to fading. Generally, neutral colors such as browns and grays are best.

Who's going to do it?

Once it has been decided what should be done to a floor or ceiling, the question arises of who is going to do the actual work.

Surprisingly, unless the work involves simple repairs or installation of a resilient tile floor or acoustical tile ceiling, the answer for most homeowners is, "I'll call up some of the firms in the yellow pages."

I find this surprising because the trend toward do-it-yourself work is so strong today, and because floor and ceiling work is for the most part very simple. The explanation, I'm sure, is that except for the manufacturers of resilient and acoustical tiles, flooring and ceiling material companies have done almost nothing to persuade the homeowner to try installing, finishing, and maintaining their products. And I can't say I have seen any change in their attitude despite the continuing surge of the do-it-yourself movement.

Be that as it may, you can take my word for it that anyone who thinks he can hang wallpaper, put up wall paneling, or lay a terrace can build or rebuild, repair, and refinish most types of floor and ceiling. True, if you mix your own concrete instead of buying a ready-mix, laying a concrete floor is arduous. And putting up a gypsum board ceiling—even with a good helper—is also arduous. But there is nothing else in the work that should give you pause.

So let's get at it.

2/ HOW FLOORS AND CEILINGS ARE BUILT

Unless you add to your house, the chances are that you will never build a floor or ceiling from scratch. Nevertheless, you should understand how they are built so you won't have trouble completing an unfinished room or rebuilding an existing floor.

Framing the first floor.

The framing for the first floor of a house consists of large beams, or girders, sills, joists, and bridging. As a rule, floors are built so the joists run across the narrower dimension of a house. In very narrow houses, the joists—each made out of a single, solid timber—extend from the foundation wall on one side to the foundation wall on the opposite side. But when the walls are so far apart that solid timbers cannot span the distance or would sag in the middle, a beam is centered between them to support the joists. In the latter case, each joist is usually made of two short timbers that overlap slightly at the beam. (A joist should not be made of two unequal lengths of timber that are spliced together partway between one of the sills and the beam. Each piece should be supported at both ends—at the sill and at the beam.)

The beam itself is notched into the foundation walls at the ends of the house so the top is flush with the sills. One or more steel or wood posts are set under it between the end walls to keep it from sagging.

Once the beam (if required) is in place, sills are laid flat on top of the foundation walls around the periphery of the house, and are anchored by bolts embedded in the masonry. The sills are made of 2- x 4-inch, 2- x 6-inch, or larger timbers. In some cases, double timbers are used.

Two-inch-thick joists are laid on edge across the sills (and beam) and toenailed to them. In normal construction, the joists are spaced 16 inches on centers (that is, 16 inches from the center of one to the center of the next), but occasionally the spacing is reduced to 12 inches or increased to 20 or 24 inches. The width of the joists depends on the distance they must span, the load they must carry (usually figured at 40 pounds per square foot), the type of wood of which they are made, and

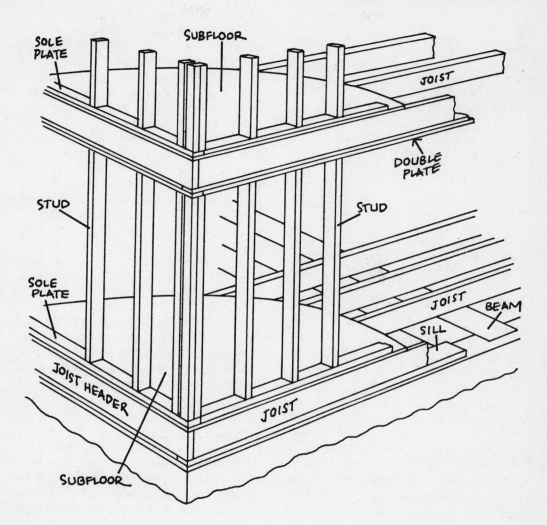

the spacing between them. For example, if the best-quality wood is used and the load to be carried is 40 pounds, 2x6s spaced 16 inches on centers will span up to 11 feet 6 inches; 2x8s, up to 15 feet 3 inches; 2x10s, up to 19 feet 2 inches; and 2x12s, up to 23 feet.

Double joists laid side by side (sometimes with a slight space between them) are used to strengthen the floor under partitions that parallel the joists. A header joist made of the same material as the joists is nailed to the ends of the joists along the sills to help hold the joists upright, to prevent wind and rain from entering the spaces between the joists, and to serve as a nailing base for the subfloor.

To stiffen the floor, hold the joists together in alignment, and transfer the load from one joist to another. Bridging is installed between all joists in straight rows no more than 8 feet apart. Most bridging—called cross

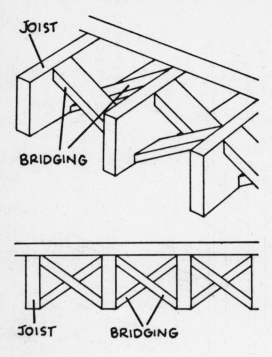

mit finish floorboards to be laid either across the joists or parallel with them. Each board was nailed to each of the joists that it crossed with a pair of 2½-inch nails.

Today, subfloors are commonly made of ½- or ⅝-inch-thick structural plywood, provided the joist spacing does not exceed 16 inches. The sheets must always be laid with the long edges perpendicular to the joists and with the short edges resting firmly on the joists. The sheets in adjacent rows are staggered so the ends do not align. Secure nailing with 2½-inch nails is essential. At the periphery of each sheet, the nails should be spaced 6 inches apart; across the intermediate joists, they should be 10 inches apart. For still greater strength and almost certain prevention of squeaks, the plywood can also be glued with a ¼-inch bead of adhesive applied to each joist.

The finish floor is laid directly on the subfloor in a new house. In an old house, however, a rigid underlayment of hardboard, particleboard, or plywood may be necessary if you're putting down a resilient floor.

bridging—is made of short pieces of 1- x 3-inch or 1- x 4-inch boards nailed between each pair of joists in an X. Prefabricated steel cross bridging is also used to some extent.

Another kind of bridging, called solid bridging or blocking, consists of short lengths of timber like that used for the joists. The ends of the blocks are squared and the blocks are nailed between adjacent joists flush with the top edges.

When the floor framing is completed, it is covered with a wood subfloor, which not only increases the strength and rigidity of the entire structure but also serves as the base for the finish floor. In the past—and sometimes even today—the subfloor was made of so-called 1-inch boards with an actual thickness of 25/32 or ¾ inch. In the best construction, these were laid at a 45° angle to the joists to brace the floor longitudinally and also to per-

Framing the second floor.

The second and all other floors in houses are built in the same way—usually with the joists running in the same direction as the first-floor joists. There are only two differences: (1) The top plates on which the joists rest—corresponding to the sills under the first-floor joists—consist of a pair of 2x4s nailed together one over the other on top of the wall studs. (2) If support for the joists is required between the outside walls of the house, it is provided—not by a beam—but by bearing partitions.

A bearing partition is an interior wall that is built like and looks exactly like a nonbearing partition. It becomes a bearing partition only because it is placed so that it supports the joists in the floor above.

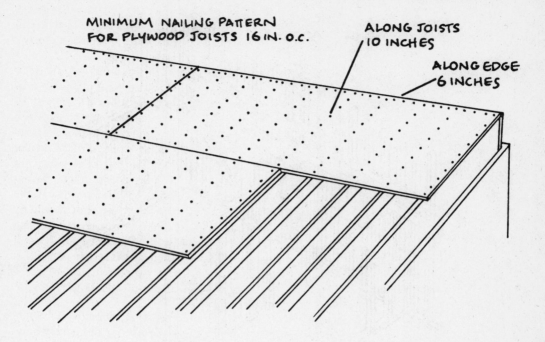

MINIMUM NAILING PATTERN FOR PLYWOOD JOISTS 16 IN. O.C.

ALONG JOISTS 10 INCHES

ALONG EDGE 6 INCHES

Bearing partitions are a frequent cause of confusion. No one can identify them just by looking at them. You must instead examine the joists above them—and this, of course, is a whole lot easier to do when the floor and wall framing are exposed than when they are covered with plaster, gypsum board, or what have you. If the partition in question is parallel to the joists, the partition is nonbearing. You can knock it out without affecting the floor above in the slightest. Here, then, is the first requirement of a bearing partition: It must be at a sharp angle (usually, but not always, a right angle) to the floor joists above.

The second requirement of a bearing partition is that it must actually support the joists in the floor above. This it obviously does if the joists end at the partition, because without the partition the joists would collapse. On the other hand, if joists continue on across a partition to another partition or to an exterior wall,

the chances are that although they may be resting on the partition they do not require it for support. True, if you took it away, they might in time sag a little, but if they were properly sized, even this wouldn't happen.

At this point you may wonder, "If it's hard to make positive identification of a bearing partition when you examine the framing, how in the world do you do it when the framing is hidden under the wall, floor and ceiling surfaces?"

The only reliable approach is to make a hole through the ceiling next to the wall you're interested in. Then poke a long, stiff wire through this, parallel with the wall. If the wire goes in several feet without striking an obstruction, it means that the joists are parallel with the wall and the wall is nonbearing. But if you strike an obstruction, and if you then turn the wire around and poke it in the opposite direction and strike another obstruc-

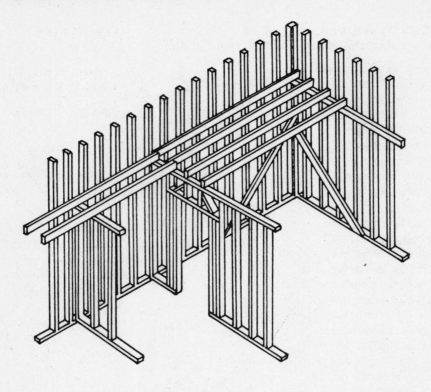

tion, it means the joists are at an angle to the wall and the wall may very possibly be a bearing wall. To settle the question, you must enlarge the hole enough so you can look into the joist space and see whether the ends of the joists are above the wall or whether the joists continue beyond the wall.

Filling gaps left in floors and ceilings by the removal of a wall.

When you tear out an interior partition, one of the steps you must take to conceal the loss is to fill the channels—which are about 3/4 inch deep and 6 inches wide—left in the floor and ceiling.

The ceiling channels are fairly easy to cope with: Just cut strips of gypsum board of the appropriate thickness to fit in the gaps, nail them to the joists, and fill the joints with gypsum board joint compound and paper tape. (See Chapter 22.)

Gaps in floors are much more difficult to fill so that they won't be eyesores. In fact, you can do it only if the finish floor is made of boards paralleling the wall removed. In this case, cut new flooring boards to fit and nail them down. But if the floorboards are at right angles to the removed wall or if the floor is made of any kind of tile, you're out of luck. About the only thing you can do—short of laying a brand-new floor—is to fill the gaps with any boards of the proper thickness, and then carpet the floor wall to wall.

When flooring materials are of different thicknesses.

All flooring materials are not, of course, of the same thickness. All the finish floors should be at the same level so they are safer to walk over, easier to clean, more attractive, and less likely to be damaged. If you're building a new house, this makes a difference in the way subfloors are constructed.

Level floors are equally desirable in an old house but are not always feasible, because no one wants to go to the trouble or expense of tearing out a perfectly good wood floor (for example) so that the new tile or vinyl he is putting down will be level with the wood in the next rooms. He just lays the tile or vinyl on top of the old wood and hopes that no one will be bothered by the slightly raised edges at the doorways. Nevertheless, when you lay a new floor, you should consider this problem and if it's at all possible to tear out the old floor rather than building upon it, you should do so.

Maintaining floor level is simple enough to figure out once you know the exact thickness of the finish flooring materials you intend to use. All you have to do is adjust the thickness of the subfloors. For example, if the subfloor throughout the house is ½-inch plywood and you are putting down a vinyl floor (approximately ⅛ inch thick) in the kitchen and strip oak floors (¾ inch), install ⅝-inch plywood underlayment under the vinyl and lay the oak directly on the ½-inch plywood subfloor. If you use ceramic tile (¼ inch thick) instead of vinyl, the thickness of the plywood underlayment should be ½ inch.

On the other hand, if you lay a brick floor next to oak, the problem becomes more difficult, because the standard brick is 2¼ inches thick. In order to align the oak with the brick, you would have to sandwich a 1½-inch layer of plywood between it and the subfloor—thus increasing the cost exorbitantly.

A more practical solution is to lower the framing under the brick by decreasing the depth of the joists 2 inches and covering them with a subfloor that is ½ inch thicker than that under the oak. To compensate for the reduced size of the joists, you should either reduce the spacing between them from 16 to 12 inches or even a little less, or double the thickness of the joists by nailing together pairs of 2-inch-thick timbers.

Cantilevering a floor.

A cantilevered floor is one that projects beyond the exterior walls of a house. Bay windows are frequently extended in this way, and it is also a fairly common practice to cantilever the entire second floor out over the front or side of New England colonial and modern houses.

If the floor joists are perpendicular to the cantilevered section, the joists are simply extended beyond the exterior wall the desired distance.

If the joists are parallel to the exterior wall, however, you must omit or remove several of

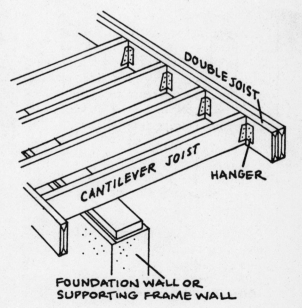

DOUBLE JOIST

CANTILEVER JOIST

HANGER

FOUNDATION WALL OR SUPPORTING FRAME WALL

the joists behind the wall and install short joists perpendicular to them. These support the cantilevered section. The illustration shows how this is done. The spacing of the short joists depends on the weight of the structure they carry. As a rule of thumb, the joists should project inward at least twice as far as they project outward.

Framing a floor for a house that is built low to the ground.

One of the problems with wood-framed floors is that they must be installed at least 6 inches above the surrounding ground so they are not exposed to moisture or easy attack by termites. This means that if you build the floor in the conventional way—atop the foundation walls—the finish floor is more than a foot above the ground. And this, in turn, adds to the total height of the house.

What do you do if you prefer a house that has the long, low look that is especially popular in the West and South? Or if you want to add a room that nestles close to the ground?

The simple answer is to build the house on a concrete slab. But if you dislike or mistrust slab construction, there is still a way to put in a wood-framed floor. Instead of making the foundation walls the same thickness from bottom to top, form a deep rectangular notch in the top of the walls around the inside edges. Install 2x4s in the bottom of the notch to serve as sills, and proceed from there as if you were building a conventional floor. See illustration.

The same approach is used if you want to build a room with a floor partially sunk below ground level. In this case, however, instead of notching the foundation walls, you should increase the thickness of the wall below the floor from the usual 8 inches to 12 inches.

Whatever the height of the floor, be sure to provide a crawl space at least 18 inches deep below the joists.

4" CONCRETE BLOCK

JOIST

SILL

8" CONCRETE BLOCKS

Framing a floor around openings.

Making an opening in a floor for a stairwell or trapdoor calls for special framing. After nailing in the joists on either side of the opening, cut four timbers from the same material to fit between them at right angles. These are the headers.

Nail one of the headers between the joists at either end of the opening. Then cut short joists (called tail joists) to extend from the headers over the nearest sills (or beams). Nail these to the headers and sills.

Now nail the remaining headers inside those you have just installed, and nail full-length joists to the outside of those framing the opening. Thus, the opening is surrounded by two thicknesses of timber.

Drilling, notching, and cutting joists.

Every time you cut a floor framing member in any way you weaken it. And since weakness is one thing that cannot be tolerated in a floor, you obviously must hold all cutting to a minimum. The precautions to follow are rather simple:

1) Avoid cutting or notching beams and girders at all costs. Only small holes are permissible—and they are not desirable.

2) Holes may be drilled through joists provided they are scattered and are at least 2 inches from an edge (preferably, they should be centered between the top and bottom edges). In 8-inch and larger joists, the holes may have a maximum diameter of 2 inches. In a 6-inch joist, maximum diameter should be 1½ inches.

3) Notches may be cut in joists only in the end third of the span. Whether made in the top or bottom of a joist, the total depth of a notch should not exceed a fourth of the depth of the joist. (That is, you can make a 1½-inch notch in a 6-inch joist.) However, the depth of a notch may equal one-third the depth of the joist as long as the center of the notch is within 8 inches of a sill or beam.

4) If you cannot avoid making a deeper notch than the basic rules permit, be sure the notch is cut in the top of the joist only. To compensate for the joist's loss of strength, either install extra joists, or support the joist on a large post topped with a 3- or 4-foot timber which will spread the load to either side of the notch.

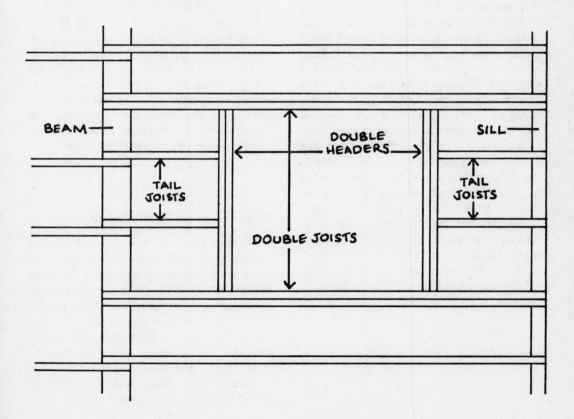

Framing a ceiling.

Ceilings are framed like floors. In fact, on most levels of a house, the ceilings and floors are carried on the same joists (which you can call floor joists if your primary interest is the floor, or ceiling joists if your primary interest is the ceiling). But even when there is no floor above a ceiling, the framing resembles that for a floor.

There may, however, be some difference in the size of the timbers required.

For example, if a plaster ceiling is hung below a floor, it's advisable to install joists one size larger than those required for the floor alone, in order to cut plaster cracking to an absolute minimum.

If the space above a ceiling is not occupied but is used only for storage (or nothing at all), the size of the joists can be reduced provided they are spaced no more than 16 inches on centers. If the best grade of lumber is used, 2x4s can be used in spans of up to 8 feet 7 inches; 2x6s in spans up to 13 feet; 2x8s in spans up to 17 feet 9 inches; and 2x10s in spans up to 22½ feet. Even if you pile 6 inches of mineral wool insulation on the ceiling between the joists, the added weight does not require an increase in the size of the joists.

Installation of ceiling surfaces is made after a house is completely closed in and the plumbing, heating, and wiring have been roughed in. The ceilings are then put in before the walls, which are followed by the finish floors.

In most cases, the finish ceiling is attached directly to the joists. The major exception is acoustical tile. This is either stapled to furring strips which are nailed to the joists or glued to gypsum backer board nailed to the joists.

Ceilings attached to the roof framing.

If, like thousands of homeowners, you have an unfinished attic, you will probably some day wonder what you must do to convert it into a bedroom. It's not the purpose of this book to go into the ramifications of the project, but I assure you that the ceiling will be the least of your worries.

The first step—if it hasn't already been taken—is to stuff 6-inch-thick insulating batts or blankets between the rafters and/or collar beams (horizontal timbers nailed across the roof, from rafter to rafter, to tie the two sides of the roof together). Use the type of insulation that is covered on one side by tough paper serving as a vapor barrier. Install the insulation with the paper facing down and fasten it in place with staples driven through the nailing flanges on both edges into the bottoms of the rafters and collar beams. The flanges should be overlapped to prevent the escape of moisture from inside the house.

Then apply the finish ceiling material over the insulation. Cover the flat part of the ceiling—under the collar beams—first, then the sloping areas under the rafters. Gypsum board, fiberboard, and plaster lath are nailed directly to the rafters and collar beams.

Post-and-beam construction.

This old, old method of construction became popular again after World War II, when architects and builders discovered it not only has several important structural and economic advantages over conventional frame construction but also makes for much greater flexibility in the design of traditional and modern homes.

In the post-and-beam system, the house is framed with a few widely spaced, big timbers rather than a network of closely spaced 2-inch

timbers. And thick planks rather than 1-inch boards may be used to make the floors, ceilings, and roof decks. The result is a sizable reduction in the number of pieces of material that must be handled; and this, in turn, tends to simplify and—hopefully—cut the cost of construction.

The system, however, is not one you can play around with. If you're building a post-and-beam house or adding on a post-and-beam room, you should have it designed by a professional; otherwise it may be structurally insufficient. Another problem the professional must overcome is the concealment of pipes, ducts, and wires.

The basic ingredients of a post-and-beam structure are 4-inch x 4-inch or larger wood posts and 4-inch-thick wood beams, or girders. In a first floor, the beams are usually supported at the ends on sills anchored to the foundation walls. Posts provide intermediate support as needed. In an upper floor, the beams are supported at the periphery of the

house by beams, and intermediate support comes from beams and conventional stud partitions. In both cases, the floors are built either of 1⅜-inch plywood with tongue-and-groove edges or of 2-inch x 6-inch tongue-and-groove planks. (Floors or roofs covered with planks are said to be of plank-and-beam construction.)

Because of the thickness and interlocking edges of the plywood panels and planks, one piece helps to support the other, and despite the wide space between beams, there is no need to install blocking under the joists. No bridging is used either. As a result, the bottom of the floor, if finished, is as attractive as the top and is often left exposed.

In a roof, the beams are usually spaced further apart, and the size is increased accordingly. In some cases, the beams run up and down the slope of the roof like ordinary rafters. These are called transverse beams. In other cases, the beams are laid across the roof. These are called longitudinal, or purlin,

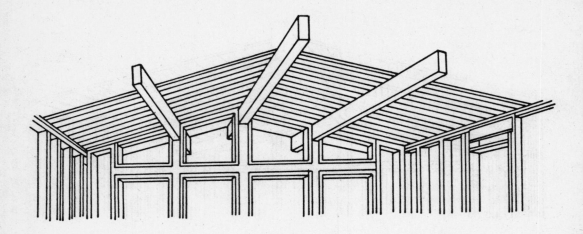

beams. The roof deck may be made of tongue-and-groove planks 2 to 4 inches thick (depending on the space between beams) or of composition board of the same thickness. Both materials are commonly left exposed and finished to serve as the ceiling. To minimize heat loss and heat gain through the roof, wood planks are covered with rigid insulation. This is unnecessary with composition boards, which have built-in insulating characterisitics. Conventional roofing materials such as shingles or built-up tar and gravel are applied over the roof deck.

Truss construction.

Trusses are giant framing members that are assembled on the ground or in a shop out of several pieces of 2-inch-thick lumber. They serve as roof rafters as well as ceiling joists. A truss for a gable roof, for example, is a large triangle crisscrossed with several short pieces of lumber to give it strength. The top, slanting members are the rafters; the long, horizontal member is the joist.

In truss construction, after the walls of a house are erected, the trusses are placed on top on 16- or 24-inch centers. Because of the way they are put together, they can span a house no more then 30 feet wide with no need for support from bearing walls. Thus, you can finish off the space underneath as one large room, which can then be divided into smaller rooms with nonbearing partitions or storage walls. The space above the ceiling joists is unusable.

Truss construction has no effect on floor construction, and its only effect on ceiling construction is to permit installation in an uninterrupted sheet over the entire space covered by the trusses. You just nail the ceiling surface directly to the bottom edges of the trusses.

Building concrete slabs.

Although I know of no statistics to verify this, more houses today are built on concrete slabs laid directly on the ground than with wood floors over basements and crawl spaces. The trend has affected not only the design of houses but also the way in which floors at ground level are built and finished.

To anyone who has always lived with wood floors, slabs are likely to be vaguely frightening. They sound cold and damp and unpleasantly hard underfoot. The truth is that they don't have to be any of these things—and if properly constructed, are not. The one real danger of slab construction is that it can—though it need not—encourage attack by termites. It is a surprising fact that termites do more damage to houses built on slabs than to those built with wood floors.

Slab construction can be used anywhere in the United States provided that the ground is stable, the water table is not close to the surface, and hydrostatic pressure is not excessive.

Three types of slab are built.

The so-called monolithic floating slab is used only on soils that are well enough drained to prevent frost damage to the slab. The slab and foundations, which are only about 1 foot deep, are poured in one piece on a well-compacted 4- to 6-inch base of crushed rock. A continuous sheet of heavy polyethylene film is inserted between the base and slab to keep moisture out of the latter. And rigid waterproof insulating boards are applied to the exposed edges of the slab to keep out cold.

Either of two kinds of suspended slab can be laid on soils subject to frost action, because it is supported on independent foundation walls which extend below the frost line. In one variety of suspended slab, the slab rests squarely on top of the foundations, so its edges are exposed (in cold climates, however, they should be covered with rigid insulation).

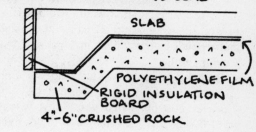

MONOLITHIC FLOATING SLAB

SLAB

POLYETHYLENE FILM
RIGID INSULATION BOARD
4"-6" CRUSHED ROCK

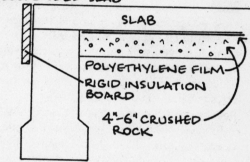

SUSPENDED SLAB

SLAB

POLYETHYLENE FILM
RIGID INSULATION BOARD
4"-6" CRUSHED ROCK

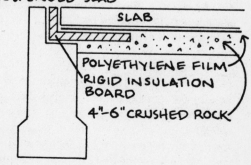

SUSPENDED SLAB

SLAB

POLYETHYLENE FILM
RIGID INSULATION BOARD
4"-6" CRUSHED ROCK

In a second variety of suspended slab, the slab rests in a rectangular notch cut in the top back edges of the foundation walls. Rigid insulation is wrapped around the edges of the slab and extends about 2 feet underneath (thus forming an L).

Both kinds of suspended slab are constructed like a floating slab on a layer of crushed rock blanketed with polyethylene film. The across-the-foundations slab gives slightly better protection against termites than the notched-in type.

For the actual details of pouring a concrete floor, see Chapter 17.

Although a concrete slab can be used as a finish floor, it usually serves only as a subfloor, and in most cases, the finish floor is glued or cemented on top of it. (Carpet is laid loose except for anchorage around the edges.) In order to provide a nailing base for a board floor, however, it is customary to cover the concrete with rows of thick wood strips to which the flooring is nailed. There are two ways to handle these to permit laying a finish floor on slabs that are built on or below grade.

In one method, the wood strips, called sleepers, consist of 1- x 2-inch boards which are saturated with wood preservative. Arrange the boards in parallel rows 16 inches on centers and embed them in ribbons of adhesive applied to the concrete. The adhesive should be either an asphalt mastic designed for bonding wood to concrete or some other type of adhesive, such as Dap Panel Adhesive or Franklin Construction Adhesive, designed for the same purpose. After the strips are stuck down, they should be further anchored with 1½-inch concrete nails spaced about 2 feet apart.

Over the strips stretch heavy (4-mil) polyethylene film. Use a continuous sheet if possible; otherwise, overlap smaller sheets on top of the wood strips. Let the film extend up the walls on all sides of the room several inches. Then nail a second course of 1- x 2-inch strips over the first. The finish flooring

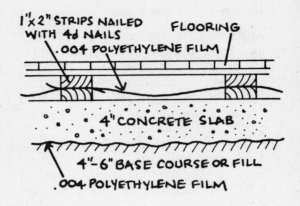

can be nailed directly to the strips, provided it is ¾ inch thick. For a more rigid floor, or if you're using thinner flooring, cover the strips with ½-inch or ⅝-inch plywood and nail the finish flooring to this.

The second method of providing a base for wood flooring is to coat the concrete, after it is

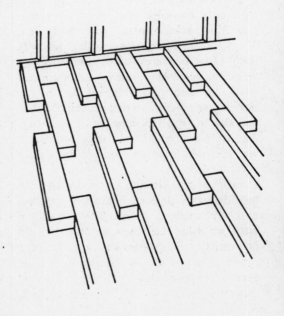

thoroughly dry, with a hot asphaltic mastic designed for bonding wood to concrete. Embed in the mastic a layer of 15-pound asphalt-saturated building felt, and allow the felt to extend upward behind the baseboards around the room. After pressing the felt smooth, apply another coat of mastic over it.

The nailing base for the flooring is made of 18- to 30-inch lengths of 2x4s that have been treated with wood preservative. Called screeds, these are laid out on the slab in staggered rows 16 inches apart. The ends of the screeds in each row should overlap at least 4 inches. Use screeds of random length so that the lapped joints in one row are not exactly opposite those in the next rows. Provide a 1-inch gap between the ends of each row and the wall to allow for expansion and contraction of the screeds. To bond the screeds to the slab, embed them in ribbons of asphalt mastic.

Finally, nail the flooring at right angles across the screeds. Or, if you prefer, you can nail the flooring to a plywood subfloor.

3/ HOW TO SOUNDPROOF FLOORS AND CEILINGS

The last apartment my elderly mother and father had was situated under an apartment occupied by a woman and a large dog. No woman or dog ever gained enemies as fast as they. From the morning after the evening my parents moved in, my father hated them—and all because they sounded as if they were walking around inside his skull.

The woman was as restless as the dog, and one seemed to infect the other. They trotted back and forth, back and forth, in their apartment—she in her high heels tap, tap, tapping on the bare wood floor, the dog's nails clicking as he trotted and making horrendous scratching noises as he came to a skidding halt at his front door when the bell sounded.

Of course, the noise that came through to my family's apartment wasn't altogether the fault of the noisemakers. The apartment was also to blame. It had been built, so to speak, of cardboard, and sounds reverberated through it as water goes through a strainer.

Such buildings are a dime a dozen today; and it makes no difference whether they are apartment houses, condominiums, duplexes,

or single-family homes. The walls and ceilings are of tissue paper; the floors, though considerably thicker, are themselves one of the principal sources of noise.

As noted in Chapter 1, noise besets modern homeowners in two ways. Some noise travels through walls, floors, and ceilings from room to room. You sit in your living room quiet as a mouse, and yet you are subjected to noises originating in the other rooms because the seemingly solid barriers around you are anything but solid.

The other kind of bothersome noise is that which bounces back at you from the walls, floors, ceilings, and furnishings when you and the other occupants of a room open your mouths.

Regrettably, the treatment for one type of noise does not control the other. This is in spite of anything you may have read to the contrary. The manufacturers of acoustical tiles tend to imply that if you cover your ceilings with their tiles your noise problems are ended. Similarly, the manufacturers of dense wall and ceiling materials tend to imply that their products will also end all your noise problems. But they are not telling the whole story. You may need both acoustical tiles and dense wall and ceiling materials in some rooms, but only one or the other in other rooms.

Rating sound-control products and methods.

There are three ways of describing the effectiveness of the products and methods used to control sound in buildings. You don't need to understand how these are arrived at; but you should have a general understanding of what they mean.

An STC (Sound Transmission Class) rating is used to indicate the effectiveness of a wall, ceiling, or floor in stopping airborne sound from traveling between rooms. The higher the number, the better the sound barrier. For example, if a ceiling has an STC rating of 25, you can readily hear and understand people talking normally on the upper side. At STC 30, loud speech can be understood fairly well. At 35, loud speech is audible but not intelligible. At 42, loud speech is only a murmur. At 45, you must strain to hear loud speech. At 50, loud speech is inaudible.

An INR (Impact Noise Rating) rating is used to express the effectiveness of a floor controlling impact noises. If the floor has a minus rating, it's considered below acceptable standards. A plus rating means it exceeds standards. The larger the number assigned, the better or worse the floor.

The effectiveness of materials in absorbing sounds within a room is indicated by an NRC (Noise Reduction Coefficient) number in the form of a decimal. The higher the number, the more absorbent the material.

Stopping noise from traveling through ceilings.

We can forget about walls at this point, for even though they transmit more noise than the horizontal surfaces of a house, they are not the concern of this book.

Contrary to much popular opinion, the way to stop noise from traveling up and down between rooms is to increase the density of ceiling surfaces. Further improvement is achieved by isolating the ceiling surfaces from the joists.

In the average low-cost development house, ceilings are commonly made of a single layer of ⅜-inch standard gypsum board nailed to the joists. Such a ceiling has next to no sound-stopping ability whatever. If the floor above is made of ½-inch plywood covered with ¾-inch oak, the STC rating for the combined floor-ceiling is only 30, whereas government studies indicate that a rating of at least 40 is required.

Increasing the ceiling thickness to ½ inch helps very slightly. Using ⅝ inch helps a little more. But you don't achieve any clearly noticeable reduction of sound until you cover the ceiling with two ½-inch layers of gypsum board or, better, with ½-inch sound-deadening fiberboard and ½-inch gypsum board. The latter construction has an STC of 42.

Another way to make a major improvement in a ceiling's sound-stopping ability is to substitute gypsum board with a special fire-resistant core for standard gypsum board. Because of its greater density, a ceiling covered with this has an STC rating about 6 points higher than a ceiling of standard board.

If you want a ceiling with a still higher STC rating, mount gypsum board—preferably of the fire-resistant type—on resilient metal channels screwed to the bottoms of the joists. The channels, sold through gypsum board dealers, are designed to separate the gypsum board from the joists by an air gap about ½ inch wide; thus, the noise that is transmitted through the joists (rather than through the joist spaces) is cut off from the ceiling surface. For how to use the channels, see Chapter 22.

Improving the sound-stopping ability of floors.

Obviously, a ceiling surface is not the sole influence over the movement of sound between the levels of a house. The floor plays a part, too, but unless you're willing to undertake some elaborate constructions, there's a limit to what you can do to make it a better sound barrier.

With a brand-new floor, one of the easier approaches is to nail 1½-inch tongue-and-groove composition board roof decking to the joists as the subfloor and apply wood flooring or carpet on top. If the finish floor is of ceramic tile or resilient material, cover the decking with ¼-inch hardboard underlayment.

Another approach if you're putting down carpet, ceramic tile, wood parquet, or resilient tile is to cover a ⅝-inch plywood subfloor with U.S. Gypsum's Mastical underlayment compound. This is troweled on the plywood (which is primed with an asphalt primer) in a ¾-inch layer and gives a semi-rough "float" finish suitable for everything except resilient tile, which requires a smooth, troweled finish. Because the material is non-combustible, it also helps to control the spread of fire.

In both of these constructions, fiberglass blanket insulation inserted between the joists adds a little further to the floor's sound resistance.

The best way to silence a floor, however, is to cover at least the area most often walked on with a soft material that deadens the sound of footsteps and moving furniture. Combined with a ceiling that stops airborne sound, such as voices, this will pretty well stop the movement of noise down through the floor as well as up through it from the rooms below.

Carpet laid on a top-quality rug cushion made of rubberized jute fibers and hairs is the number one choice. The thickness of the carpet makes little difference. A thick carpet—provided it is densely woven—is desirable only when a rug cushion is omitted.

Other flooring materials that muffle impact sounds are, in order of preference, cork, vinyl, and linoleum.

Reducing noise transmission through existing floors and ceilings.

Because most of the floor and ceiling treatments described require that the joists be exposed before you start covering them, they are best suited to new houses and new floors. What, you may wonder, is the most practical way to control noise through an existing floor and ceiling?

There are two steps you can take:
1) Increase the density of the existing ceiling—whatever that may be—by covering it with a new layer of ½-inch standard gypsum board or, better, ½-inch fire-resistant gypsum board. For maximum results, the new board should be glued to the ceiling rather than nailed. This is possible, however, only if the ceiling is level and smooth.
2) If the gypsum board does not sufficiently reduce the noise of footsteps on the floor, you should cover the floor with carpet, cork, vinyl, or linoleum.

Plugging holes that leak sound.

Amazing as it may seem, a small hole in a ceiling, floor, or wall sometimes transmits as much sound as the entire surrounding surface.

It follows that if you go to the trouble of soundproofing a floor or ceiling, you should take pains to look for and seal openings through which unwanted sound will rush.

You are most likely to find such openings around electrical boxes, heating and air conditioning outlets, and radiator and water pipes. To plug them, simply squeeze in a special acoustical caulking compound sold by gypsum board manufacturers. Since the caulking is not paintable, it must be carefully applied so that it is hidden by the covers or plates for the electrical boxes, or anything else that might hide it.

Deadening noise within a room.

It's much easier to bring quiet to a room that reverberates with the noise made by its occupants than to stop the transmission of sound between rooms. Sometimes this can be accomplished simply by putting down wall-to-wall carpet over a rug cushion, covering a wall with a heavy wall hanging, or hanging larger draperies at the windows. But the most common practice—undoubtedly inspired by the determined promotion of manufacturers—is to apply acoustical tiles to the ceiling.

Most homeowners make one mistake in doing this: They buy the tiles that look the prettiest without giving enough thought to their sound-absorption characteristics. Thus, the result is sometimes a room that is either still noisier than desired or is so quiet that it has a muffled, dead feeling.

Always check the noise reduction coefficient of tiles first.

As a general rule, the proper tiles for really noisy rooms such as kitchens and family rooms are those with an NRC of .60 to .70. In other rooms, an NRC of .40 to .50 is usually adequate.

The area where the tile will be installed should also be considered, because if you don't put in enough tiles you won't reduce noise sufficiently, and if you put in too many you'll reduce noise to such an extent that conversation becomes difficult and unpleasant. But here, unfortunately, you are flying blind unless you hire an acoustical engineer—not just an acoustical contractor—to make a careful study of the room. Very few people do this.

What, then, should you do?

The cautious approach—which is also the wise approach if the room you're treating is not excessively noisy—is to start out by installing in the center of the ceiling a rectangle of tiles equal to about two-thirds or three-quarters of the total ceiling area. Later, if this does not lower the noise level enough, you can easily extend the installation.

But the usual practice—which is based on nothing but hunch—is to cover the entire ceiling at the outset and hope that it works out satisfactorily. Generally it does. If you are not so lucky—if the room turns out to be too quiet—some of the tiles can be removed.

4/ REPAIRING FLOORS AND CEILINGS

Just yesterday I began work on the gypsum board ceiling in our laundry. It had developed a bad bulge in one spot as a result of an old leak in the bathroom above. Once before, when the bulge was smaller, I had tried to "remove" it by concealing it under a spreading coat of gypsum board joint compound (more on this later). But that hadn't worked because the weakened board continued to sag even after it had dried out. So when Elizabeth and I decided a week ago to redo the entire laundry, I knew I would have to excise the bulge and insert a patch. I didn't anticipate any trouble. But I ran into it anyway.

As soon as I cut the bulge open I discovered that the gypsum board had been installed over an old plaster ceiling. Worse, when the gypsum board had been put up, a large area of the plaster had been removed, and in order to make the gypsum board level over the entire ceiling, a new joist which was about ¾ inch deeper than the old joists had been installed. I was therefore faced with the task not only of removing the weighty, sagging plaster above the bulge, but also of furring down the old joists to the level of the new one.

This may sound as if the job turned into a nightmare. Actually, it took me only a couple of hours (instead of the thirty minutes I had expected). But the incident makes a point: When you go to work on the ceilings in an old house, you can't be certain what problems you will encounter.

Tom Rooney, my oldest son-in-law, has had more than his share of trouble with old-house ceilings. Tom is a scientist who grew up in the city without ever taking a hammer in hand. As a matter of fact, he once gave solemn oath to my daughter that he wasn't a do-it-yourselfer and never—no, never—intended to be. Famous last words. Since they bought an ancient house in the Boston area a couple of years ago, he's been building and rebuilding nonstop.

The ceilings in the house were a mess. Made of plaster over wood lath, they were cracked in every direction. Not tiny map cracks that you can usually conceal with paint, but gaping cracks. Nevertheless, when I was given my first tour of the house, I opined that patching plaster and gypsum board joint compound would take care of them. But when

26

Tom and I got up on ladders for a closer look, I backed off from my hasty long-distance judgment. Although the plaster wasn't actually sagging, many of the cracked sections had partially broken away from the lath, and they were hanging in place by the grace of God.

"Well," I said, "I guess you'll have to knock the ceilings out and put up gypsum board."

Tom was clearly worried, and I had a pretty good idea why. He mistrusted his ability to install a good-looking ceiling; also, he didn't feel he had the time to give to such a project.

Some days later, he was presented with a much simpler, though far less attractive and durable, solution. "Why don't you try ceiling buttons?" his paint dealer suggested.

"Ceiling buttons!" Tom said. "You must be kidding."

"Not a bit. Take a look at this," the dealer countered, and rummaging around, he came up with a metal button about the size of a nickel. It was perforated and slightly concave. "All you do is drill a few little holes through the loose pieces of plaster, put a screw through the center hole in this button, and drive the screw into the wood lath until the button pulls the plaster up tight against it."

"Works just like a washer," Tom said.

"Same thing, only the button's thinner and has holes in it so you can smear gypsum board joint compound over it in a thin coat to hide it."

"But won't that make a hump on the ceiling?" Tom asked. "Lots of humps—one over each button."

"No, you just put on enough joint compound, feathering the edges away from the buttons, until the surface looks smooth."

"I'll try 'em," Tom said.

The buttons worked like a charm. To be sure, the ceilings don't look like new ceilings if you examine them closely, but they are no longer threatening to fall down, and they are smooth enough to satisfy anyone except a perfectionist. That's the first point in the buttons' favor.

The second is that they saved Tom an enormous amount of work—at least for the time being. But I don't want to lay too much stress on this idea, because the fact is that even with the ceiling buttons Tom had a great deal to do.

That's how ceilings are: When they call for repairs, don't think you're going to get the work done in a minute. It may not be difficult, but it's a pain in the neck.

Floors are no better. But whereas ceiling repairs are slow because of the awkward position you must work in, floor repairs take time because (1) the source of trouble is often concealed and (2) the cure cannot be hastened.

The problems and repairs covered in this chapter are of a general nature and are not limited to any one particular type of floor or ceiling. Other problems and repairs are covered in the chapters following.

Creaking floors.

The noisy floor is probably one of the most common problems that homeowners must contend with. It is also one of the knottiest because it is difficult to pinpoint the source of the noise and then to silence it.

The first step is to examine the offending floor, because sometimes, just by walking on it, you can feel and see exactly what is causing the noise. However, I must confess that most cures are effected by trial-and-error procedure, because you can't be sure whether the noise emanates from the finish flooring, the subfloor, or the framing under everything.

Even though examination of a floor may not reveal the course of action to take, it does at least give you an idea about which of the several possible cures you should attempt and in what order. Understandably enough, most people start with the likeliest cure and work down. My own usual—though not invariable—procedure is to start with the cures that will not affect the appearance of the floor and work down to those that will. Do as you wish. Any one of the following steps may do the trick:

1) If it's possible to get at the underside of the floor, look for gaps between the subfloor and joists and drive thin wedges (pieces of wood shingle are excellent) into these. Be careful not to open the cracks any wider than they already are.

2) Working on the underside of the floor, drive 1-inch roundhead screws up through the subfloor into the finish floor to stop noisy friction between the two. Fit washers over the screwheads before driving the screws.

3) Squirt lubricant into the joints between wood boards or blocks. I always used to use powdered graphite, even though it made a black mess that had to be washed off several times with paint thinner, until I came across a white greasy powder called Dry-lube. This works just as well as graphite and is much neater. Silicone lubricant in a spray can is another possibility.

4) Fasten down squeaky floorboards or subflooring by driving 3- or 4-inch galvanized finishing nails diagonally through them into the subfloor and/or joists. To prevent the nails from splitting hardwood flooring, drill small holes partway through the flooring first. Countersink the nailheads and cover them with plastic wood or, simply, floor wax.

5) If diagonal nails don't hold, try long screws.

Sagging floors.

The first thing to note here is that there's a difference between a sagging floor and a floor that sags because the house is settling. It's usually fairly simple to differentiate between the two because the latter runs downhill toward a foundation wall, and unless the slope has been caused by the rotting of the sill (which you can quickly determine), it is attributable to the settlement of the foundation wall. There's nothing you can do about this unless you hire a contractor to jack up the entire side of the house, including the foundation wall, or to jack up the aboveground portion of the house and add new timbers between it and the foundations.

An ordinary sagging floor, on the other hand, has simply developed a low spot somewhere because the joists weren't properly sized or spaced or because they were weakened by insects, decay, or old age. Jacking up this kind of floor is a job you can do yourself—if anyone at all can do it.

What you need is a jack post or jack screw. They're essentially the same thing except that posts range from about 3 to 9 feet in length while screws are only about 1 to 2 feet. Buy whichever you need: The cost is low, and you'll need it for a longer time than you can afford to rent.

Position the post under the joist nearest the center of the sagging area. To spread the pressure exerted downward and upward by the post, stand it on a pair of 2x4s or 2x6s placed one on the other on the basement floor, and place another pair of timbers between the top of the post and the joists. The timbers should be about 4 feet long. Set the upper pair at right angles to the joists.

Screw the post up tight against the sagging floor; then tighten a little more to raise the floor ¼ inch. Leave it alone for at least a week to allow the wood to adjust to its new position, then raise it another ¼ inch. And continue this way until the floor is finally more or less

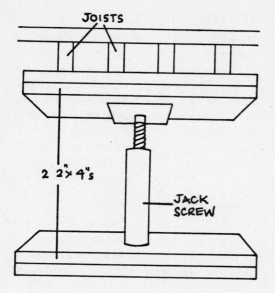

If you can get at the underside of the floor, take a hard look at it while someone tramps around on top. Stretch a string tightly across the bottom of the joists, so that it touches them, to help you detect any up-and-down motion in them. If you notice any, your best course of action is to remove the bridging and slip new joists of the same size between the existing joists. Fasten them to the sills (or beams) at either end. Then drive 4- or 5-inch galvanized finishing nails diagonally down through the floor and subfloor into them. Finally, cut blocks from the joist lumber and nail them between each adjacent pair of joists in rows 8 feet apart.

If you can't detect any movement in the joists when someone is walking overhead, install one or two additional rows of crossbridging or solid bridging between the joists.

level. You can then replace the jack post with one or more wood posts or Lally columns. If you use the former, 4-inch x 4-inch timbers should be adequate, but 6x6s are preferable if the sagging area is very large and weighted under very heavy furniture. Lally columns are circular steel posts filled with concrete. Use a 4-inch size.

Large holes in a wood floor.

The holes may have been cut in the floor for registers. Repair is easy if you can get at the underside of the floor. Just cut a pair of 2x4s to fit between the joists, and nail them into place on either side of the hole. They should extend for half their thickness into the hole. Nail 1x2s under them to the sides of the joists

Vibrating floors.

When a floor bounces or shakes under the weight of people walking, it means either that the joists are too small or too widely spaced or that the bridging is inadequate. Unfortunately, unless you can see the joists, bridging, and underside of the floor, there's no way of telling what the problem is or of doing anything about it. In other words, if the floor that vibrates has a ceiling underneath, you can't fix it short of tearing out the ceiling or supporting it on a couple of unsightly posts.

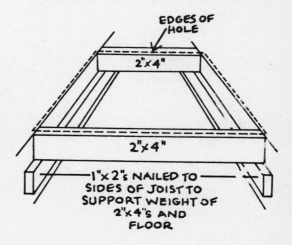

so they won't break loose under weight. Then cut a piece of plywood the thickness of the subfloor to fill the hole and nail it to the top of the blocks. Lay finish flooring over this.

If you can't get under the floor, you must cut the flooring and subfloor back as necessary to the sides of the joists. To cut along the joists, drill holes through the floor next to them and saw out the wood with a keyhole saw. Then nail 4-inch boards to the sides of the joists tight against the bottom of the subfloor, and nail the plywood patch over these.

Weak spots in a floor.

If a weak spot doesn't show signs of caving in and you can get under the floor, nail several 2-inch x 4-inch cross blocks between the joists under the spot. If the subfloor is rotten, however, you should cut out the floor and subfloor in a rectangle and treat it like a large hole (see above).

Gaps between floors and shoe moldings.

These appear because the shoe moldings are nailed to the baseboards and cannot subside as the floors settle. To close such a gap, pry off the molding and pull out the nails point first (so they don't leave holes in the face of the molding). Scrape off paint that may prevent the molding from fitting tightly against the floor and baseboard. Then nail the molding into place with 3-inch finishing nails driven at a 45° angle into the joint between the floor and baseboard. Since the molding isn't nailed to either the floor or the baseboard, it will no longer develop gaps.

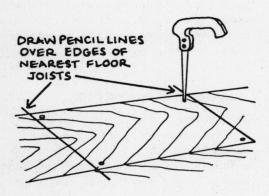

DRAW PENCIL LINES OVER EDGES OF NEAREST FLOOR JOISTS

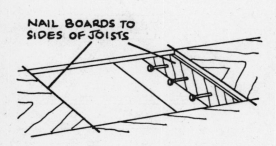

NAIL BOARDS TO SIDES OF JOISTS

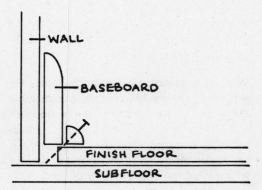

WALL

BASEBOARD

FINISH FLOOR

SUBFLOOR

Gaps between shoe moldings and baseboards.

These result from the fact that the moldings are nailed to the floor so that they pull away from the baseboards when shrinkage occurs. Correct the problem as above.

Shoe moldings badly dented and splintered.

The only thing you can do is to replace them with new moldings. Ideally you should use ⅝-inch quarter-rounds, but these seem to have disappeared from most lumberyards, so you will probably have to settle for either ½- or ¾-inch quarter-rounds. The first may be too small to cover wide cracks under baseboards. The second are too big to be attractive. But this doesn't seem to worry millwork manufacturers.

If you are going to paint the moldings, give them a priming coat first. Then cut them to fit the wall spaces and nail them in with 3-inch finishing nails driven diagonally into the joint between the floor and baseboards. Countersink the heads and cover with spackle. Then apply a finishing coat of paint.

The job is easy, but the moldings must be cut carefully so they form neat joints at the corners. Use miter joints at outside corners (an outside corner is shaped like the knuckles on your hand when the fingers are bent), coped joints at inside corners (like the bend inside your bent fingers). Use a miter box to make miter joints.

In a coped joint, one piece of molding is cut straight across and jammed into the corner. Then the end of the other piece is cut to a quarter-round outline and fitted against the side of the first piece. To cut this outline, you should miter the second piece as if it were to be used in a miter joint. Then, with a coping saw, cut through the molding, following the front edge and removing the 45° miter.

Interior thresholds badly worn.

Thresholds are beveled boards that are nailed to a floor directly under and parallel to the bottom of a door. They are rarely used today except under exterior doors, but are common in older houses under interior as well as exterior doors. Their purpose is to minimize drafts and, sometimes, to conceal an ugly joint between the flooring in one room and the next.

To remove a worn interior threshold, pry off the stops on the doorjambs. (Stops are the thin strips of wood against which a door bears when shut.) Then pry up the threshold, taking pains not to scar the floor or the jambs. If you find that the flooring continues right under the threshold, there is no need to replace the strip. Just fill the nail holes and refinish the floor to match the exposed areas on either side.

If you put in a new threshold, saw it to fit snugly between the jambs and nail it down with finishing nails. Countersink the heads and cover them with plastic wood stained to match the finish floor. Apply new finish, and replace the stops.

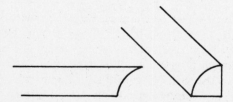

Sloping or uneven ceilings.

The best way to level a ceiling is to tear it out and build down anew from the exposed joists. This obviously makes work—not to mention a hideous mess—but you save materials and you don't reduce the height of the room.

When the ceiling is removed, stretch strings taut and level across the joists to delineate the low spots and high spots. Then, starting from the lowest joist and progressing toward the highest, fur the ceiling down to the strings by nailing 1- x 2-inch or 1- x 3-inch wood strips across the joists on 16-inch centers. Nail the first furring strip directly to the joists. As you move toward the high point in the ceiling, insert increasingly thicker shims, or wedges, between the furring strips and joists. The shims must be long enough to be nailed securely to the joists so the furring strips can, in turn, be nailed securely to them. Finally, when the furring is in place, cover it with ⅜- or ½-inch gypsum board (see Chapter 22).

If you don't want to bother with removing the old ceiling, nail the furring strips over it into the joists. Install the strips in the same way—at right angles to the joists.

(Note that the furring strips can be nailed along the joists rather than across them, but as you reach the high point of the ceiling, the strips become so thick that they are difficult to nail securely and do not provide a sound, rigid nailing base for the gypsum board. Furthermore, you use much more lumber for the strips and shims than you need for a cross-the-joist installation.)

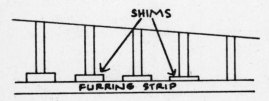

Ceiling bulges.

As I said at the start of this chapter, the first thing I did to repair my bulging laundry ceiling was to smear on gypsum board joint compound. I realize that it sounds like a crazy solution—the more so since the bulge grew more pronounced. But one of the best—and certainly the simplest—ways to obliterate bulges in ceilings and walls is to enlarge them until they blend into the background and disappear from view (though not from reality).

The approach works as well on plaster as on gypsum board. In fact, it works on any smooth material. And if you want to use spackle or patching plaster rather than gypsum board joint compound, you may (but the joint compound is a great deal easier to work with).

The whole idea behind the operation is to apply joint compound around the sides of the bulge until the bulge's sharp outline disappears. To do this, you just keep adding more and more joint compound day after day and spreading it out over the surrounding ceiling further and further. Sand thoroughly between coats, and when you are finally satisfied that the bulge is no longer visible, give the area a final careful sanding to remove all unevenness in the surface.

Ceiling paint failures.

Some paint problems are attributable to the material to which the paint is applied. For example, paint on a plaster ceiling may fail if it was applied before the plaster had cured sufficiently. Similarly, paint on a board ceiling may develop ugly brown stains where the resin in the wood has bled through. Such problems, though not very common, are touched on in the following chapters.

The most common of all ceiling paint problems is the one that afflicts all kinds of painted ceilings—cracking and flaking. The usual cause is dampness, but I am beginning to wonder if there isn't some inherent weakness in at least certain latex paints that causes them to start flaking badly even after they have been on a ceiling for several years. I've had several recent sad experiences that lead me to this conclusion, but in the absence of proof, there is no reason to belabor the point.

When ceiling paint starts to crack and flake, the first thing you should do is to find and stop whatever leaks there may be in the roof or floors above. The only thing you can then do—whether there are leaks or not—is to remove the damaged paint. This is a difficult, tedious job you would hardly wish on your worst enemy. Scraping in an upside-down position is ineffective. Sanding—even with a reciprocating electric sander—doesn't do the job thoroughly. When my entire living room ceiling went to pieces for no apparent reason after it had held up beautifully for two years, I came to the reluctant conclusion that the only way to get the stuff off was to use either an electric or chemical paint remover. Normally, I favor the former for big jobs, but in this case I had to take off latex paint, and heat doesn't work well on latex. So I bought a couple of gallons of a paste-type paint remover and set to. Surprisingly, the work proved easier than I had expected. On the cracked and flaking areas the chemical was so potent that the paint peeled off and fell to the floor without much assistance. It was only on the sound areas that I had to use a scraper.

If there are any secrets to the operation, they are pretty simple. Apply the paint remover—paste-type only; that must be emphasized—liberally to a 6- to 9-square-foot area. Let it cook for about ten minutes. (You soon discover how large an area you should cover and how long to let the remover work.) Then scrape off the paint with a broad knife or wallpaper scraper (I favor the latter because it's sharper and stiffer). Be sure to get off every fleck, otherwise it will show

through the new paint. Finally, wash the ceiling with a rag dipped in mineral spirits to neutralize whatever remover remains. (This is theoretically unnecessary with water-wash paint removers, but why take chances?)

For how to repaint the ceiling, see Chapter 5.

Mildew on ceilings.

If the mail I receive from all around the country asking about various problems is indicative, almost every homeowner except in the dry Southwest is plagued by mildew on ceilings and walls. Getting rid of the pest—temporarily—is no problem. Just wash the surface with a strong solution of chlorine bleach. Or if the ceilings are very dirty as well as mildewed, wash with a solution of 2/3 cup trisodium phosphate and 1 quart chlorine bleach in 3 quarts of warm water.

Putting a permanent stop to mildew is another matter, however.

If the ceiling needs to be repainted, either buy paint containing a mildewcide or add a mildew-retardant (available from the paint store) to whatever paint you use.

But the only reasonably sure way to keep mildew under control is to ventilate the house well and frequently and to turn the heat up now and then during the summer. Installing a small ventilating fan in each bathroom—where mildew principally occurs—is highly advisable.

In addition, you should take steps to hold down excess humidity in the house. These include covering the soil in crawl spaces with a continuous sheet of heavy polyethylene film or with a 2-inch slab of concrete; installing ventilated openings in crawl spaces; and reducing your use of water by such measures as using a damp rather than a wet mop to clean kitchen and bathroom floors, covering all pots when cooking, taking shorter showers and baths in somewhat cooler water, and so forth.

5/ FINISHING AND DECORATING FLOORS AND CEILINGS

The time will come when the combination of carpet and resilient flooring usurps wood's place as the number one flooring material, but until then it is safe to say that since most existing residential floors are made of wood, most residential floors need to be finished and refinished. And no one can argue the point that virtually all ceilings also need finishing and refinishing.

For a great many homeowners such a prospect holds, if not dismay, at least a feeling of resignation. Even though it isn't difficult, the work is time-consuming. And after a year or two, all your effort appears wasted when you survey the scratches on the floors and the film of grime on the ceilings.

I have no bright words of comfort to offer. There are no new miracle finishes that change the picture. On the contrary, the worst problems we have with floors and ceilings—and probably always will have—are caused by the finishes we put on them.

34

Finishes for wood floors.

I suppose it's a sad commentary on my intelligence, but the fact remains that when I started refinishing floors almost forty years ago, I selected the worst possible finish rather than the best. From this I progressed to the second worst, and then the third, and it wasn't until comparatively recent years that I finally came to the best finish and stuck with it.

Advertising led me astray. It hardly occurred to me that the worst finishes are promoted heavily, and that the best are promoted hardly at all. In fact, when I went into the largest paint store in my area a couple of years ago and asked for a gallon of penetrating floor sealer, the proprietor looked me straight in the eye and said, "They don't make that stuff any more." I've since discovered that, despite the size and success of the store, the people who run it know next to nothing about paint. Or maybe they put too much faith in the promotions they're exposed to.

Anyway, I won't bother you with the no-good floor finishes—only the best. Of these, there is really only one, but I must mention the other since it is currently receiving considerable publicity. That is urethane or, more properly, polyurethane varnish. The toughest of all varnishes, it is said to be the clear finish supreme for wood. But that's going a little too far. Admittedly, it gives an excellent glossy finish if you apply two or three coats. But it doesn't hold up forever under traffic, and when it finally does begin to look threadbare, you can't touch it up without showing the same kind of lap marks that plague the users of other kinds of varnish.

Penetrating wood floor sealer is clearly the standout finish for all kinds of wood floor.

Because it penetrates the top surface of the wood, it becomes, in effect, a part of the wood and wears away with it. Furthermore, because of its penetration, it seals the pores of the wood against dirt and stains.

Although the surface film left by a penetrating sealer shows scratches, these are less objectionable than those in finishes that lie entirely on the surface, and they are readily obliterated by sanding and the application of a touch-up coat.

The finish is semiglossy and doesn't change the color of wood drastically (unless you use a tinted sealer). When you apply a touch-up coat on a worn spot, the new sealer amalgamates with the old so that there is no visible difference between the touched-up spot and the surrounding area.

Finally, the finish is not slippery. It's resistant to water. It takes and holds wax well. And it's easy to maintain.

Brands that are approved by the Maple Flooring Manufacturers Association on the basis of laboratory tests are as follows. These are as good for oak and other woods as for maple.

Absolute Heavy Duty Finish - Absolute Coatings, Inc., 34 Industrial St., Bronx, N.Y. 10461

Bal-Dur Imperial Primer Sealer - Balding Products, Inc., 28 Main St., Geneseo, N.Y. 14454

Bal-Dur Imperial U Primer Sealer - Balding Products, Inc.

Dura Seal Penetrating Sealer - Minwax Co., 72 Oak St., Clifton, N.J. 07014

Franklin Chekit Penetrating Seal - Purex Corp., Carson, Calif. 90745

Penetrating Triple XXX Seal-O-San - Huntington Laboratories, Inc., P.O. Box 710, Huntington, Ind. 46750

RPM Penetrating Floor Seal - Republic Powdered Metals, 2628 Pearl Rd., Medina, Ohio 44256

Triumph Wood Seal - Western Chemical Co., 417 S. Fourth St., St. Joseph, Mo. 64501

Floor paints are less durable than most of the clear finishes, because they lie entirely on the surface and therefore have less resistance to scuffing and because they *show* wear to a greater extent. In six years I have painted the fir floor in our laundry three times, and despite careful preparation and use of the best oil-base floor enamel I can find, the wear in the main traffic areas becomes mildly objectionable within a year of application.

If you're starting fresh with a well-sanded floor, you'll do better with an epoxy primer followed by a coat of epoxy enamel. If your floor has already been painted with something else, however, epoxy can't be used unless you remove the old paint first. Since this entails considerable work, you may prefer—as I do—to use an oil-base enamel until the built-up coating reaches the point where it simply must be taken off. Then switch to epoxy.

Finishes for masonry floors.

Concrete floors seem to hold paint even less well than wood. If the floors are the slightest bit damp, of course, the paint flakes off like the bark on a sycamore tree. But even on a dry floor, it doesn't wear well in traffic lanes.

Epoxy is the only thing to use. I am forever amazed at how much tougher it is than even the best of the oil-base enamels and new latex paints. If you etch the clean concrete with muriatic acid, rinse thoroughly, and then apply an epoxy primer and one or two coats of epoxy floor enamel, the floor should stay attractive for three or four years or more because it is extremely resistant to abrasion and moisture (even a little subsurface moisture—but don't count on its holding if the concrete is resting directly on the ground without a moisture barrier underneath). Its one serious drawback is that it is slick as glass when damp—but after all, you don't expect floors in a house to be damp very often.

Epoxy will also do better than anything else on brick or stone if you ever decide that a floor of these materials needs a face change. Not many homeowners could bring themselves even to consider such a rash act, however, and there's no reason why they should unless the natural color of the floor is hideous or unless they have allowed it to become stained beyond redemption.

Needless to say, stains on brick or stone are not inevitable but they do occur. It's for this reason that floors made of them should be finished with a transparent, colorless masonry sealer, which penetrates the pores and produces an almost invisible surface film. Two coats give nearly complete protection.

Finishes for ceilings.

Ceilings of wood or plywood don't require a finish. In time, however, they will probably acquire a certain amount of grime, and the only way you can then clean them is by sanding—a laborious operation. To avoid this eventuality, you may prefer to give ceilings in living and sleeping areas two coats of clear wallpaper lacquer. The result does not change the wood noticeably but does permit you to wash the ceiling when dirty.

You can use lacquer in kitchens and bathrooms, too. But because kitchen ceilings need almost annual washing to remove grease, and bathroom ceilings need washing now and then to remove mildew and soil deposited by the damp atmosphere, a tougher finish is advisable. Varnish is the best selection. True, even the clearest varnishes darken wood, but you can scrub them pretty much to your heart's content.

Varnishes are available in dull, semigloss, and gloss finishes. From the appearance standpoint, the dull is best, but it is not quite as easy to clean or as durable as the others. Gloss is exactly the reverse: very cleanable and durable, but so shiny you will hate it. That leaves semigloss as a compromise.

But paint is the favorite ceiling finish. I don't believe I have seen anything else used on plaster, gypsum board, or acoustical tiles, and it's often used on wood, too. The kind you use depends on the ceiling location. As I just said, kitchen and bathroom ceilings need fairly regular washing; others generally do not.

Since alkyd paint is more washable than latex, and semigloss finishes are more washable than flats (again I rule out gloss finishes, which are too shiny for use on such large areas as ceilings), an alkyd semigloss should be used in kitchens and bathrooms. Use latex elsewhere—simply because it's easier to apply and requires less after-application cleanup. But on wood ceilings—regardless of location—always use alkyd. Latex just doesn't do very well on wood, despite claims to the contrary.

How to finish a floor.

Because the all-important preparatory work for this job varies, it is covered in later chapters on specific types of flooring.

The actual painting (I use the word loosely to include application of other types of finish) should be done after everything else in the room has been painted. The one exception is when you are using a clear finish, in which case—unless you have a very steady hand—it's best to finish the shoe moldings at the foot of the baseboards after the floor is finished. This makes it easier to wipe off any paint that drips from the moldings: It would penetrate the pores of the bare flooring otherwise.

Before starting to paint, map out your application strategy so you don't paint yourself into a corner. Do the closets and alcoves first, and then work across the main body of the floor toward your exit.

If there are heavy pieces of furniture or appliances you can't easily move out of the room, pull them into the center of the floor and paint wide borders around the floor first. If possible, let the finish dry a couple of days before you push the pieces back into place against the walls. To help protect the newly finished floor, put a tarpaulin or old mattress pad under the furniture and appliances, and slide them on that.

Apply the finish with a wide brush or short-napped roller. The latter speeds the operation greatly and is almost essential with epoxy paint, which is difficult to brush out. Since rollers used in anything other than a water-base paint are hard to clean for use on a later project, buy a cheap roller cover rather than a good one, and throw it away after the

floor is done. Between the first and second coats, place the cover in your roller pan and pour in enough solvent to cover the bottom third or half of the cover. Roll the cover back and forth several times until it is saturated. It should then remain soft so you can use it again the next day.

Whether using a brush or roller, apply the paint—especially the first coat—in one direction, and then go back over it in the other direction. This not only helps to fill pores that might otherwise escape if you work only in one direction but also helps to remove puddles and to produce a smoother, more even coat.

Let each coat dry completely before walking on the floor. The drying time usually printed on the paint-can label gives only a general idea of how long drying takes. If the room is cold or the humidity high, actual drying usually takes longer.

Light sanding or steel-wooling between coats is necessary, as a rule, only with penetrating sealers and varnish. After sanding, vacuum the floor thoroughly.

How to apply a spatter finish.

For a spatter finish, the floor is first painted a solid color or given a hard transparent finish, then it is spattered or polka-dotted with paint of another color or colors.

For a very fine spattered effect, apply the paint with a spray gun. For a coarser, bolder effect, load the tip of your paintbrush with paint. Hold a stick of wood in your left hand, and bring the ferrule of the paintbrush down sharply against the stick. This causes the paint to spatter on the floor in random dots and spots. The size of the dots depends on how full you have loaded the brush and on how close the brush is to the floor when you strike the stick.

How to apply a marbleized finish.

Using a piece of marble as your guide, practice making this finish before starting on the floor.

Apply a solid base coat first and let it dry. Over this apply a second color in squiggly lines with the end of a worn feather. Blur the paint here and there. After this dries, apply a third color in the same way. If you prefer an even greater blurred effect, apply the third coat while the second is wet.

How to stencil a floor.

Stenciling a painted design on a floor is like stenciling a piece of furniture. Buy the stencils, or make your own by drawing designs on stiff, oil-treated cardboard and cutting them out with a razor blade.

Place the stencil in position on the floor and stick down the edges with masking tape. Pour a little paint into a saucer and tap the bristles of a stencil brush in it; then transfer the paint to the floor by tapping straight down on the cutout part of the stencil. Leave the stencil in place until the paint becomes tacky; if you lift it too soon, the paint will seep under the edges.

How to finish ceilings.

The ceiling is the first part of a room to be painted. Cover the floor with drop cloths or newspapers, and if you leave the room at any time, wipe your shoes off carefully so you don't track paint with you.

Although there's a temptation to work from a plank suspended between stepladders when painting a high ceiling, it's better not to do so, because it is too easy to back off the plank when working over your head. Even though it's a nuisance to move a stepladder, stand on that. The alternative if you are using a roller is to stand on the floor and put an extension handle on the roller. But you may find this harder on the arm and back muscles than you suspect.

A roller is undoubtedly the easiest tool for painting ceilings because it covers a large area in a short time. The faintly pebbled surface that it leaves is not noticeable. The texture of the roller cover used depends on the ceiling texture. A short-napped roller is correct for a smooth ceiling. For a textured ceiling, select a somewhat longer nap—about ⅜ inch.

The first step in painting a ceiling is to do a narrow strip around the borders. This is called cutting in. You can go all the way around the ceiling at one time or keep just ahead of work in the middle of the room. Use a brush, a small cutting-in roller, or edging pad. I prefer the brush because I can work it all the way into the crease between ceiling and walls and into corners. With a roller or pad, you're less able to do this. On the other hand, if you use a roller to paint the center of the ceiling, the texture of the brushed-on paint around the edges doesn't match that of the rolled-on. The only way to minimize this mismatch is to make the brushed strip only about 1 inch wide and then to go over it—as close as possible to the walls—with your large ceiling roller.

The proper way to work on a ceiling—particularly a sizable ceiling—is to apply the paint in strips across the narrow dimension. Start at one end of the room and work toward the other. As soon as you finish one strip, return to the wall at which you started and apply another parallel strip. Make the strips 2 feet wide if you're using fast-drying latex paint, 3 feet wide for slower-drying alkyd. This procedure is designed to eliminate noticeable lap marks between strips. If the strips are narrow and short, the first strip should still be wet when you start the next strip, and you can even out the paint where the strips overlap.

To avoid splattering, don't overload a roller. Roll the paint out in any direction other than that in which you will finish rolling. Start the first stroke in a dry area and work into the adjacent wet area. After applying two or three roller loads in this way, go back over the entire area in one direction. Don't press down too hard, and take pains to roll out the ridges of paint that may be left at the ends of the roller, because they won't disappear by themselves.

Handle a paintbrush the same way: Don't overload it. Apply the paint any which way, but always smooth it out with light strokes in one direction.

Latex paint should completely dry within two hours, so you can apply a second coat the same day. Wallpaper lacquer applied to wood can also be recoated the same day. But other finishes should be allowed to dry overnight.

Sanded and textured ceiling paints.

These are widely used by developers because they help to cover up imperfections in the sloppy taping jobs their crews do on gypsum board ceilings. You can use them, too, to cover up cracks, pits, and bumps in both gypsum board and plaster ceilings. I happen to dislike them (1) because they are usually a sign that there's something wrong with the ceiling base, (2) because they're hard to wash, and (3) because the sand-finished ceilings I inherited in my present house flake badly for no apparent reason.

Sanded paints are latex formulations with fine sand suspended in them. You can make your own simply by adding fine, clean sand—like that used in swimming pool filters—or a synthetic sand called Perltex to any ordinary paint. The paint is usually applied with semicircular strokes of a special texturing brush.

Textured paints are very thick latex or oil-base paints used to produce a special texture. The kind of texture depends on the application method. First the paint is brushed on the ceiling in an even coat; then it is textured with a crumpled newspaper, sponge, whisk broom, comb, or anything else producing a pattern. There are also special texturing rollers.

But these paints have drawbacks. In addition to being hard to wash (because of the texture), they cover only a small area, which adds to the cost, and they are fiendishly hard to remove or smooth out if you tire of the texture. (This is also true—but to a lesser extent—of sanded paints.) The only ways to cope with the paints are to scrape and sand, scrape and sand, until you drop in your tracks; apply paste-type paint remover and scrape; or apply repeated skim coats of gypsum board joint compound until the hollows are brought up to the level of the high points.

If you're applying a sanded or textured paint on a light-colored surface, one coat is usually sufficient. To change a ceiling from dark to light, however, you may need two coats. Test first on a small area. If one coat doesn't give enough coverage, use an ordinary paint of the same type and brand for the first coat. Apply the sanded or textured paint only as the second coat.

Wallpapering ceilings.

The idea of covering a ceiling with wallpaper or vinyl wall covering strikes terror into many people, but it's not as difficult as you may think—especially if the ceiling is low enough so you can reach it without climbing on a ladder.

A ceiling is always papered before the walls, which are usually being papered, too, because instead of trimming off the paper at the corners between the ceilings and walls you should let it extend 1 inch down the walls.

Paper across the room, not lengthwise, so

you don't have to cope with extremely long strips. Cut the strips to the width of the ceiling plus 2 inches.

After you've decided which end of the room you will start papering from, measure out from the wall the width of the paper minus 1 inch. Make a chalk line across the ceiling at this point.

The adhesive for hanging ceiling paper is like that for wallpaper but somewhat stiffer so the paper won't peel off of its own weight. Brush the adhesive on the back of each strip just before you hang it. Fold the ends of the strip in toward the center. As you carry the strip and paste it to the ceiling, support it on a roll of wallpaper held in your left hand (or right hand if you are left-handed).

When hanging the paper, face the starting wall. Hang the paper from right to left if you are right-handed, from left to right if left-handed. Pull open the folded strip, align the edge with the chalk line, and press the paper to the ceiling so it overhangs the side wall by 1 inch. Smooth it down with your smoothing brush. Then open out the remainder of the strip and smooth it on the ceiling.

Hang succeeding strips in the same way. Butt the edges.

Wallpapering floors.

Every once in a while I hear that some decorator has applied wallpaper to a floor. It's a pretty far-out idea, practical only if the floor is made of smooth-surfaced plywood, hardboard, or particleboard, or of wood strips that have lost all tendency to expand and contract and that are sanded smooth and level. But if you find a wallpaper that has just the right color and pattern for a floor covering, why not?

Use a vinyl wall covering rather than paper. It is far more resistant to wear and the hard, frequent cleaning that will be called for.

Run the wall covering in the same direction as the floorboards: It looks better if joints between boards (which are bound to show through the vinyl) parallel the pattern. If the pattern is a strong one, start papering from the middle of the room toward the side walls so the strips next to the walls will be of equal width. If the pattern is indefinite, however, you can start papering from the wall inside the entrance door and work straight across to the opposite wall.

Use the adhesive recommended for the particular type of wall covering. Take great care to eliminate all bubbles in the wall covering when smoothing it on the floor, and to stick the edges (which should be butted) down tight.

Vinyl wall covering needs no finishing or waxing. If you use wallpaper, cover it with two coats of urethane varnish when the adhesive is totally dry.

Ceiling beams.

Just because your house was not built with posts and beams is no reason why you shouldn't have a beamed ceiling in your family room, living room, or elsewhere if you think it would add to the appeal and character of the house. Urethane beams textured and colored to resemble oak have been on the market for a number of years and have proved to be a hit with homeowners everywhere.

Though the beams are made in relatively short (about 6- to 8-foot) lengths, they are fashioned so they can be put together end to end, without visible seams, to span wide ceilings. Because of their light weight, they can be fastened to any sound surface with adhesive (and a few finishing nails to hold them while the adhesive sets). Just cut them and put them up.

If you prefer larger or more finished-looking beams, you can make them yourself out of 1-inch pine boards. These are called boxed beams because they are nothing more than rectangular troughs. To make one, mark the inside width of the beam on the ceiling and nail up 1-inch x 2-inch furring strips just inside the lines. Then nail the boards forming the sides of the beam to the strips, and nail a third board between them about ¼ inch up from the bottom edges. Paint or stain the beams as you like.

6/ **WOOD FLOORING-STRIP**

Wood won its position as the number one flooring material because it was, long ago, the only widely available, easily worked material, and because it was the logical thing to use (in buildings that were built up off the ground as most were). It hangs on to the top spot today primarily because it is beautiful and durable, but of course it has other points in its favor. It's adaptable to any style of architecture. It's comfortable and strong. It's relatively easy to maintain, and when it does become badly worn, it can be restored to pristine condition. And it is not only reasonably economical initially, but also becomes extraordinarily economical over the long run because it lasts for generations.

Just why and when narrow flooring boards—called strip flooring—became the standard type of wood flooring I can only guess. The date was some time in the 1800s. The reasons probably were that the civilized world started running out of the huge trees from which wide planks could be cut and that the advent of steam power made it possible to turn out small sizes of lumber at a much faster clip than water power could produce large

42

sizes. Whatever the actual facts, when home builders today decide to put in wood floors, they almost automatically think of strip flooring. It isn't the prettiest choice but it is the most adaptable and practical. For about $75, a do-it-yourselfer can cover 100 square feet with No. 1 common oak, and he can do it with little difficulty, though hardly at breakneck speed.

Types of wood.

You can build a floor of just about any species of wood except the softest and most splintery. But you must first find a reliable supply of the wood you like and you must then have it milled to your specifications. This costs money. In the long run, you'll be better off to stick to what's available. The list includes the following:

Oak. In usage, oak ranks far ahead of all other woods as a flooring material. Its open, coarse, but interesting grain makes it easy to identify, and when people entering a house see it, they instantly associate it with quality.

Of the two kinds of oak used for flooring, red is more common than white, but the only discernible—and important—difference between them is in their color. Red oak has a slight reddish or pinkish cast; white oak has a brownish cast. The two are usually sold and installed separately except in the lowest grades.

Both red and white oak are graded in the same way. The first breakdown is between quarter-sawed and plain-sawed stock. The former consists of boards that have been sawed so that the growth rings form an angle of 45° or more with the faces of the boards. When nailed to a subfloor, the boards have a close, rather straight grain which looks as if it might have been produced with a comb. Plain-sawed stock, on the other hand, is ripped out of a log in such a way that the growth rings form a shallow angle with the faces. This produces an irregular grain pattern not only in each board but also from one board to the other.

Quarter-sawed stock is superior in both appearance and durability, so naturally you pay a premium for it. It is available in only two grades—clear, meaning that the boards have virtually no imperfections, and select, which is just a notch below clear.

Plain-sawed oak is available in five grades. Clear and select are the best. Then come No. 1 common, No. 2 common, and 1½-foot shorts (a mixture of very short pieces of all the higher grades).

Maple. Maple flooring cut from sugar maple trees—not from other maple varieties—has outstanding wearing qualities, and for this reason it is the favorite wood flooring for gymnasiums and other buildings where it is subjected to abuse. It has never been widely used in homes, however. This is perhaps because it is more durable than you need in a home, but I suspect the main complaint against it is that it's not very handsome. The grain is fine and indistinct; the color ranges

from almost white to medium brown. Because of the grain and color, imperfections tend to stand out. On the other hand, you don't have to worry about the imperfections if you buy the first grade. It's only in the second, third, and fourth grades that you find them.

Beech. Beech is much like maple: very hard, strong, and durable, with a fine, close grain and color ranging from almost white to pale brown. It's available in first, second, third, and fourth grades.

Birch. This is another wood in the same class as maple and beech, and graded in the same way. Its color and texture are somewhat more distinctive, however.

Pecan. Pecan is a heavy, strong, extremely durable type of hickory that has become particularly popular for paneling and cabinets. It also makes a fine flooring, ranging from reddish-brown to creamy white and characterized by large pores. The grades are first grade, first grade red (which is the same as first grade but all boards are cut from the dark-colored heartwood), first grade white (also like first grade but with all boards cut from the light-colored sapwood), second grade and second grade red, and third grade.

Teak. Teak is so highly prized around the world that, even though it is not a very scarce wood, it commands a very high price. No readily available wood flooring costs more, but no other will give you more to boast about. Teak is exceptionally strong, heavy, and durable. It is a handsome medium brown and has a strong grain, or pattern, much like black walnut.

One problem with teak is that it is oily, and this poses finishing problems. But these can be largely avoided if you apply two or three coats of penetrating sealer or moisture-cured urethane varnish.

Walnut, cherry, ash, and hickory. There is very little call for these woods in strip flooring; consequently manufacturers have not attempted to classify or grade them. But some manufacturers do make something of a specialty of them. All are hard and durable. Walnut and cherry are particularly striking in appearance.

Yellow pine. Cut from several species of southern pine trees, yellow pine is a strong, hard, resinous, yellowish wood with little beauty—especially when cut flat-grain. Usually used only for basic structural purposes, such as framing, sheathing, and subflooring, it is nevertheless a rather common choice for finish flooring when economy is of prime importance—in simple vacation houses, for example. The best flooring is graded B & B. Lesser grades are C, C & Btr, D, and No. 2.

White fir. Although white fir is a softwood, it has better resistance to abrasion than most softwoods and is therefore suitable for flooring where you want economy, and appearance is of secondary importance. This is not to say it is an unattractive wood. A light, soft shade of brown, it acquires a pleasant reddish tinge when finished. The boards are very uniform in color and appearance and have a straight, close grain.

Cypress. A southern softwood, cypress also makes a good wear-resistant floor ranging in color from cream to a pretty brownish-red. It has outstanding resistance to decay and termites. Like other softwoods, the strips are available in 3-, 4-, 5-, and 6-inch widths. You can also get 2-inch x 4-inch and 2-inch x 6-inch flooring. All come in select grades as well as No. 1 and No. 2 common.

Prefinished flooring.

To relieve builders and do-it-yourselfers of the need to sand and finish a wood floor once it has been laid, many flooring manufacturers sell flooring that is prefinished at the factory. The arguments in favor of these new materials are persuasive. For one thing, since the finish is applied under controlled conditions and by methods that cannot be used in the field, it is more uniform and durable than the average home-applied finish. For another thing, the floor is ready for service as soon as it's laid.

A rare advantage (because this flooring is not available from all manufacturers) is that you can sometimes buy strips with chamfered (usually called beveled) edges and ends so that each piece stands out distinctly from the others—like the planks in a plank floor.

There are, however, disadvantages which the advertising doesn't mention:

The cost of the flooring is high.

The flooring must be put down with extreme care, because any scratches or hammer marks you make cannot be obliterated except by almost complete refinishing of the damaged strip.

Also, since the flooring should be laid after all other work in the room is done (in order to protect the flooring), you must provide space under baseboards, door trim, cabinets, and so on when they are installed so that the flooring can be slipped in underneath them. This takes time and also creates problems when you fit in the final flooring strips.

Types and sizes of strip flooring.

Most of the strip flooring laid in homes is tongued and grooved not only on the sides but also at the ends. (The flooring is said to be side- and end-matched.) This is preferred because it produces snug joints, simplifies nailing, and increases the solidity and strength of the floor. However, you can buy square-edged, or jointed, flooring.

Sizes vary somewhat not only among wood species but also among types of flooring. Generally, however, the thicknesses of strips range from 5/16 to 55/32 inch, widths range from 1 to 3½ inches.

The thicknesses in most common use in houses are ⅜, ½, and ¾ inch. (Until very recently, ¾-inch flooring was actually 25/32 inch thick.) The most common widths are 1½, 2, 2¼, and 3¼ inch. The single most popular size is ¾ x 2¼ inch, and because this is produced in greatest volume, it costs proportionally less than other sizes.

Estimating how much flooring you need.

Strip flooring is sold in bundles by the board foot. Each bundle contains 24 board feet. The first step in figuring how much flooring you should order is to make an accurate measurement of the room, including closets and offsets. Multiply the length by the width to give the total area in square feet. Multiply your total by the percentage given below for the size of flooring you will use. Then add the two figures to find the total number of board feet required.

55%	for	¾ x 1½ inch flooring
42½%		¾ x 2 inch
38 1/3%		¾ x 2¼ inch
29%		¾ x 3¼ inch
38 1/3%		⅜ x 1½ inch

30%		⅜ x 2 inch
38 1/3%		½ x 1½ inch
30%		½ x 2 inch

For example, if a room measures 10½ x 16 feet and you are putting down ¾- x 2¼-inch strips, you should order 168 square feet plus 64 square feet (.383 x 168)—a total of 232 board feet, or ten bundles.

Laying a strip floor over a plywood or diagonal-board subfloor.

I remember the time a friend of mine who had had very little building experience decided to lay an oak floor in a semifinished attic room. Soon after the flooring was delivered from the lumberyard, he called me in distress:

"You should see this stuff," he exclaimed. "It's all bits and pieces of different length. I think I've been had."

"You're OK," I assured him. "That's the way board flooring comes—in a lot of different lengths."

"Yeah, but if I'd known that. . . .Look how much more work it's going to be to put down."

"In one way, yes," I answered. "You have to sort pieces out, arrange them end to end. But in another way it's easier because if you ever tried to fit one long strip into another, you'd find it's difficult to get a snug fit all the way along the boards."

"But it's going to make a ratty-looking floor—a jumble—like a jigsaw puzzle."

"Take a look at your present floors," I suggested. "They're made the same way, and did you ever think they looked bad?"

He wasn't totally mollified, but when I dropped by a few days later to see how he was getting along, he was as pleased as punch, not only with the floor but also with himself. "Looks pretty good, doesn't it?" he said.

"But I'm still not sure why flooring comes that way."

There are reasons, but they don't really matter.

Wood flooring should be laid in a room or house after all other work except the installation of trim has been completed. If you have put down a concrete slab or plastered the walls or ceilings, let them dry for a couple of weeks before the flooring is even delivered.

Delivery of the flooring should be made three or four days before you actually start work, in order to give the wood time to adjust its moisture content to that of the room. Turn on the heat in the room to about 70°. As soon as the bundles of flooring arrive, open them and pile the strips loosely on the floor.

While you're waiting for the strips to become conditioned, check the subfloor once more to make sure it is nailed down securely. (If you do this now, you'll avoid squeaks later on.) Then sweep the floor clean and cover it with 15-pound asphalt-saturated building paper. Lap the edges of the strips 4 inches. In the area directly over the heating plant, use a double-weight paper. It's also a good idea to put insulation in the ceiling above the furnace to prevent excess heat from opening cracks between the floorboards.

To give the house a unified look, the new flooring should be laid in the same direction as that in the adjoining rooms. If the room is isolated, however, the strips should run lengthwise in the room.

Before nailing down any strips, lay out several rows loosely on the subfloor and shuffle them around to get the best possible combination of colors, grains, and end joints. Discard any strips that are warped, or cut out the straight sections in them for use in closets. All joints in adjacent rows should be staggered at least 6 inches. Use the shortest strips in closets and in the middle of the floor where they'll be covered by a rug, the longest lengths at entrances and other prominent locations. When a strip must be cut to fill out the end of a row, use the leftover piece to start the next row.

If you're laying a floor in just one room, start at one side of the room and work systematically across to the opposite side. Align the first boards carefully with the wall or, if the wall is crooked, to a chalk line snapped on the floor*. Provide a ½-inch space between the boards and the wall to allow for expansion. (A similar space is necessary along all other walls.)

Lay the first strips with the grooves facing the wall, and face-nail the grooves to the subfloor. (Face-nailing means to drive nails through the exposed face of a board.) The nails should be placed so they will be concealed by the shoe moldings.

Then blind-nail the tongue edge. All other boards except those in the last row at the opposite wall are blind-nailed only. In blind-nailing, drive the nails through the tongues of the boards at a 45° to 50° angle. (The nail points should be set in the crease between the tongue and the shoulder of the boards.) Drive the nails down as far as possible without striking the flooring, then countersink them with a nail set.

You can also put a floor down with a hammer-actuated power nailer available from rental outlets. This saves time but adds to cost.

The best nails for laying wood flooring are screw-type flooring nails with spiral threads.

---

* To strike a chalk line on a floor or ceiling, tie a stout cotton cord between two nails and run a piece of blue chalk along the cord. Hammer the nails into proper position with the cord stretched taut between them. Grasp the cord in the middle, lift it straight—perpendicular to the floor; not at an angle—and let it snap on the floor.

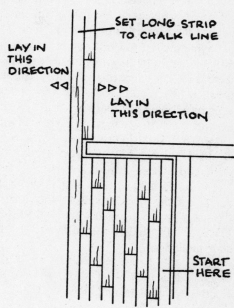

Old-fashioned, wedge-shaped cut nails may also be used, and casing nails, which resemble finishing nails, are sometimes used for thin flooring. For ¾-inch flooring use 2¼-inch (7-penny) or 2½-inch (8-penny) nails. These are ordinarily spaced 10 to 12 inches apart, but when the subfloor is made of ½-inch plywood, one nail should be driven into each joist and an additional nail should be centered between joists. For ½-inch flooring, use 1¾-inch (5-penny) nails spaced 10 inches apart, and for ⅜-inch flooring, use 1½-inch (4-penny) nails spaced 8 inches apart.

After the first row of flooring strips has been fastened down, the others must be snugged up against them before nailing. To lock the joints, put a scrap piece of flooring against the exposed edge of the piece being placed and rap it with your hammer. Never hammer the flooring boards themselves because that will damage the tongues. Lock the end joints in the same way. If, after nailing down a row of boards, you think the joint between it and the preceding row is too wide, go back along the row with a wood block and tap the strips into better position.

When flooring extends through a doorway, a long strip that runs continuously from the room through the doorway into the adjoining space should be installed against the wall projecting between the room and the space. The illustration shows the procedure to follow. Lay the flooring in the room up to the doorway. Then lay the extended strip through the doorway parallel to a chalk line struck on the floor on the other side of the doorway. After

nailing the extended strip in the area beyond the doorway, lay the flooring from it in both directions.

If a room has a step down across one end or at a doorway, the sequence in which the flooring is laid must be changed to some extent because the strip forming the front part of the "tread" should be at least 3 inches wide. If you are laying 3¼-inch flooring, use a piece of this and plane off the tongue. If your flooring is narrower than 3½ inches, buy a 4-inch board matching the flooring. In either case, round off the exposed edge of the board to match the treads on a stair. Then face-nail the board across the step lengthwise. It should overhang the riser beneath it by 1 inch.

If the flooring runs parallel to the tread, start laying the strips from the tread and work across to the opposite wall. If the flooring is perpendicular to the tread, however, start laying it in the normal way from one of the walls.

Then, when you reach the tread, cut the strips off at right angles so they will butt tightly against the back edge of the tread.

Laying the last row of flooring is the most difficult part of putting down a floor because there is so little space to work in. If the space between the preceding row and the wall is less than the width of the flooring plus the ½-inch expansion space, you must rip the boards with a saw. To close the joint between the two rows of boards, place a block of wood against the wall, slip a large screwdriver between the block and the tongue-edge of the last strips, and wedge the strips into place. Then anchor the strips by driving nails through the faces. Place the nails so they will be covered by the shoe moldings.

After the floor is down, install the baseboards a fraction of an inch above the flooring. The shoe moldings are then laid directly on the flooring and nailed.

Laying strip flooring over sleepers or screeds.

Follow the procedure outlined even though a subfloor is omitted. Nail the flooring to every sleeper or screed, and wherever a strip extends over lapped screeds, nail it to both.

End joints between flooring strips need not fall over the supporting timber because the tongue-and-groove joints prevent the strips from sagging. (If you're using square-edged flooring, all end joints must be supported.) However, end joints in adjacent rows of boards must not break over the same space between sleepers or screeds.

Laying strip flooring over an old finish floor.

When old wood floors are in such bad condition that you cannot salvage them by heavy sanding, the only thing you can do is to put down a new floor. This is done either by ripping up the old floor and laying the new one on the subfloor, or laying the new floor directly over the old. Both approaches present a few problems. Taking up an old floor is an unmitigated chore. On the other hand, if you save the old floor, you may have to level it—also a chore; you must saw down the doors in the room; and the finish floor will be higher than that in adjoining rooms.

If you remove the existing floor, go over the subfloor foot by foot; make sure it's nailed down tight, and drive down all protruding nails. Then cover the subfloor with 15-pound building paper, and carry on as if you were building a brand-new floor.

If you decide to cover the existing floor, do it with 5/16-, ⅜-, or ½-inch-thick strips. First remove the doors and shoe moldings. Nail down the floorboards securely. If any of them are badly buckled, try flattening them with large screws; otherwise, chisel or plane them down.

If an occasional board is sunken or broken, don't wory about it: As long as the new boards are at right angles to the old, they will bridge the gaps. However, if the new boards are laid parallel to the old, the gaps must be filled with strips of wood or latex cement. On the other hand, if a sizable area of the old floor is depressed, it makes no difference how the new one is laid. The low area must be brought up to level by troweling in latex cement.

Cover the floor with building paper and then follow the directions above for laying a brand-new floor. The new strips should be perpendicular to the old except in very narrow rooms. In the latter case, position the new strips so the joints between rows do not fall directly over the joints between the old boards.

At entrances to adjoining rooms, bevel the edges of the new strips so you won't stub your toe when walking between rooms. Where the strips are perpendicular to the doorway, nail a beveled strip across the ends.

How to sand a wood floor.

As they come from the factory, flooring strips have such a fine finish that you'll probably find yourself wondering from time to time, "Will I really have to sand the floor when it's completed?" The answer, I'm sorry to say, is yes, and you'll soon see why. Despite the smoothness of the strips, there is just enough difference in their thickness to produce an unevenness in the finished floor that can be removed only by sanding.

So that's why new strip floors need sanding. Old floors, or course, need sanding to remove old finishes and/or make the floors smooth and level again. Both kinds of floor are treated in the same way.

Two machines must be rented for the job: a drum sander, for sanding the center of the floor, and an edger, or disk sander, for sanding the edges. The former is a big, heavy machine resembling an upright vacuum cleaner which you control by a handle as you walk behind. The sandpaper is mounted on a steel drum which revolves from the top toward the front so that it pulls the machine forward and produces a sanded strip with a straight grain. Some models operate on 120 volts, others on 240.

The edger is a much smaller and lighter 120-volt machine operated from a standing or kneeling position. Sandpaper is clamped onto a flat disk that cuts in a circular pattern like the sanding attachment for an electric drill.

The two machines together rent for roughly $12 a day. Unless you run into grave prob-

lems in refinishing an old floor, you shouldn't need the machines for more than one day per room, and with experience, you should be able to do two rooms in a day.

The rental agency will also sell you the necessary sandpaper. Three types are needed for each machine: coarse, medium, and fine. You'll probably be jolted by the cost of each sheet, but take heart: It lasts a lot longer than ordinary sandpaper. Buy four to six sheets of each grade for each machine. Any you don't use can be returned for full credit.

Before leaving the rental agency, make sure that the cloth dust-collection bags on the sanders are sound and clean, and that you have been given the tools for mounting the sandpaper. It's also a good idea to buy a few extra 20-amp fuses in the event that the drum sander overloads the circuit into which it's plugged.

The only other tool needed is a small hand scraper to get into the corners the sanders can't reach. The best type is one holding a narrow steel cutting strip in a clamp at the forward end of a straight handle.

To prepare for sanding, remove everything that's loose from the room. Don't overlook the pictures on the walls. Open the windows and close the doors. The sanding dust can make a fearful mess if the collection bag doesn't pick up as it should.

Remove the shoe moldings so you can sand closer to the baseboards. This will also protect the moldings from damage by the edger. Also remove registers, electric plates, and metal patches nailed over mouseholes. If possible, loosen radiators so you can at least move them out of the way enough to sand under them.

Nail down all loose, protruding, or creaking floorboards. Countersink the heads of exposed nails and screws. Finally, sweep the floor clean, and sweep it frequently throughout actual sanding operations so you have a clear view of what you've done and what needs to be done.

The most important rule in floor sanding is never to pause in one spot without lifting the sander off the floor. And before you turn off the machine, lift it. If you don't, the sander will dig right through to China before you know it.

Start with the coarsest sandpaper necessary to do the job and progress in sequence down to the finest. Always use the fine paper for the last cut even if you don't think the floor needs a final polishing. On a new floor and some smooth old floors, you can usually start with medium paper. But if a floor is very rough or uneven, use coarse paper first.

Always operate the drum sander lengthwise on the flooring strips—with the grain. Work across the grain—but at no more than a 45° angle—only if the floor is extremely uneven or if many of the boards are cupped upward or downward.

Do the center of the floor first with the drum sander. The procedure is to make a straight pass forward the length of the room, back up over the same strip, then move the machine over and make an adjoining forward and backward pass. Don't overlap strips too much.

When the center of the floor is sanded, run the edger all around the edges of the floor. It doesn't make any difference whether you move constantly in one direction or go back and forth over a difficult area. But don't stop, and don't tilt the machine. The disk must always rest flat on the floor; otherwise, it will carve out deep circles you can't eliminate.

When you've sanded every area possible, use the hand scraper to clean corners, around pipes, and in other difficult spots. The secret to success is in keeping the blade sharp. This is done by turning it upside down and running a fine-toothed file over it lengthwise. Since the blade is set at a slight angle in the handle, hold the face of the file at right angles to the blade; then the back edge of the blade will be perfectly square, and as you pull it, it will

shave off the top of the wood as neatly as a razor blade. However, even the sharpest scraper is ineffective unless you bear down on it. This requires two hands, as a rule. Pull the scraper toward you with one hand while pushing down on the head with the other. Always scrape with the grain.

Once the floor is clean and smooth, touch up any rough spots remaining by hand-sanding with medium-fine paper. Then vacuum the entire room—not just the floor but also the walls, windows, doors, and woodwork—to get up every speck of dust.

Bleaching.

In an old floor there may be some stains that sanding cannot remove. These must be bleached in order to get rid of them. You can buy prepared bleach at a paint store, or make your own by dissolving ½ cup of oxalic acid crystals in 1 quart of water. Swab this liberally on the stains; let it dry, then apply more until the stains disappear.

Neutralize the acid by brushing on a solution of 1 cup of borax in 1 quart of hot water. Finally, when the wood is totally dry, sand it smooth. (Water raises the grain.)

Finishing the floor.

If the prefinished flooring on the market is any indication (and it must be, because manufacturers rarely bring out products for which there isn't some indicated demand), there is a considerable trend today toward wood floors of a darker than natural color. If you share this feeling, your first step after sanding and bleaching a floor is to stain it. Use a clear oil stain unless the maker of the final floor finish you use specifies a water stain.

Oil stain is easy to work with. The only possible mistake you can make is to let it penetrate so long that it leaves a slightly darker color than wanted. To avoid this, test it on part of the floor in a well-lighted closet. Brush it on and let it stand for five minutes. Then wipe off the excess with a rag and examine the color. If too light, brush on another coat and let stand for two minutes. Continue thus until you learn how long the stain should stand to give the color desired. Follow this timing when you stain the entire floor.

Apply the stain with a wide brush in an even coat. My practice is to start at one end of the room and work the long way of the boards to the other end. I cover an area about 30 inches wide, which is as far as I can reach comfortably when on hands and knees. I make the strip as straight as possible along the front edge so that there will be a minimum overlap when I put down the next strip; thus there is less chance of getting a darker color in overlapped areas.

After the stain has penetrated for the proper length of time, go over the strip with clean, dry rags to sop up the excess stain and even out what remains.

When the entire floor is stained, let it dry for twenty-four hours. Then apply the final finish (see Chapter 5).

Maintaining wood floors.

No matter how tough a floor finish, it should be waxed for extra protection against soiling, scratching, and staining. Use only a solvent-base wax, not a water-base type like that for resilient floors.

On a newly finished floor, use either a paste wax or a buffable liquid wax and apply two thin coats (much more durable than a single thick coat).

Paste wax is a chore to apply because you must get down on your knees and rub it on the floor with a damp cloth. After it has dried for about twenty minutes, buff it by hand or—better—with an electric floor polisher. From then on, whenever it begins to look dull, you need only to buff it to bring back the warm glow. Because of the wax's excellent wearability, you don't have to apply another coat—except in heaviest traffic areas—for six months or longer.

Buffable liquid wax is somewhat less durable. It is applied like a water-base kitchen wax with a wool applicator. After it's dry, buff it. Thereafter, buff whenever it looks tired.

After the initial application of whichever wax you use, don't apply any more than is absolutely necessary, because you just build up a thick film that hides the wood and absorbs dirt. Occasional buffing gives better results. There comes a time, however, when a new coat of wax is needed. You can then apply the wax you originally put on or use a solvent-base clean-and-polish wax. In either case, one coat is ample.

All paste and buffable liquid waxes contain a cleaning agent, but the special clean-and-polish waxes are more thorough. They are also self-polishing. Applied with a wool applicator, they do not require buffing when dry. This obviously makes them easier to use, but unfortunately, they do not last as long.

When the wax buildup on the floor finally becomes too great or the embedded dirt becomes too objectionable, the floor should be "stripped" clean. Use one of the waterless floor cleaners available. Then start all over again with two thin coats of paste or buffable liquid wax.

Outside of waxing, the only thing a wood floor needs to maintain its appearance is frequent dusting or vacuuming. The best bit of floor-care advice I ever ran across is this:

Vacuum a wood floor when you vacuum the carpet.

Wax a wood floor when you shampoo the carpet.

Strip and wax a wood floor when you replace the carpet.

The same kind of maintenance is required on prefinished floors. The one step that may be omitted is the application of the initial coats of wax—but only if the flooring manufacturer waxed the flooring as well as finishing it.

Repairing wood floors.

Scratches in a clear finish. The scratched-up floor is the most common floor problem and one of the hardest to cope with even when the scratches have no more substance than a cobweb. If I allowed myself to get upset about scratches, I think I'd be on my knees at least once a month struggling with them. Even as it is, Elizabeth presses me into duty about every six months.

Finding the proper cure for a scratched wood floor is complicated by the fact that sometimes the scratches are only in the finish and sometimes they go through into the wood, but without a magnifying glass, it's often difficult to tell one kind from the other.

If scratches are in the finish only, the lazy way to deal with them is to rub on a little wax and buff hard. In fact, this is the only way you can treat scratches in varnish, because if you make any attempt to touch them up with new varnish or to sand the entire area and coat with varnish, you end up in worse trouble than before you started: You just cannot touch up varnish without creating dark areas where the new varnish overlaps the old.

If the scratches are in a floor finished with penetrating sealer or shellac, however, you have a choice of treatments. One is simply to wax and buff. The other is to remove the wax in the scratched area by scrubbing with benzine or a special floor cleaner of the type used to strip wood floors. Tone down the scratches as much as possible with medium steel wool or medium-fine sandpaper. Then brush on a thin new coat of sealer (if the floor is sealed) or white shellac (if the floor is shellacked). Because the new finish softens the old, the two blend together without leaving lap marks.

Scratches that go through the finish into the wood can be completely removed only by sanding until there are shallow troughs in place of the original jagged lines. Maybe these are an improvement, but I question whether they are enough of one to justify the effort. I prefer a cure that is only 25 percent effective but requires less than 10 percent as much energy. If the floor was stained originally, touch up the scratch with a brush dipped in the stain and partially dried on a newspaper. (If you flood the scratch with stain, the concentration will leave a dark line as noticeable as the line of bare wood.) Then apply a coat of finish. If the floor was not stained, simply run some of the finish into the scratch with a small artist's brush and hope for the best.

Scratches in paint. One of the few advantages that painted floors have over those with a clear finish is that scratches and other flaws can be completely concealed with minimum effort. Sand them lightly to remove loose paint along the edges. Fill with spackle, and when dry, sand smooth with medium sandpaper wrapped around a block of wood so you don't remove the spackle from the center of the scratch. Then apply two coats of floor enamel.

Water spots. Water spots generally leave white marks which disappear when you rub them with cigarette ashes and vegetable oil. If this doesn't work, daub the spots with spirits of camphor and let dry for half an hour. Then rub with a rag dipped in oil and then in rottenstone. Clean off the residue with benzine. Then apply wax.

Black spots caused by water dripping from radiators, puppy accidents, or wet diapers are much more difficult to remove. In fact, you'll be lucky if you make any progress at all. But you can't admit defeat until you've tried the following steps in order: (1) Rub the spot well with coarse steel wool and a good floor cleaner or paint thinner. (2) Sand with medium and then fine sandpaper. (3) Bleach with oxalic acid and make at least three applications before calling it quits.

When one of these cures works, even out the finish around the spots with fine sandpaper and apply new finish and wax.

Alcohol spots. Rub with a rag dipped in vegetable oil and rottenstone. Rinse with benzine and apply wax.

Rust stains. Mix 1 part sodium citrate with 6 parts water, and then mix with an equal portion of glycerine. Add whiting and stir to a thick paste. Spread this on the stains and let stand until it dries in two or three days. Then scrape off. If the stains remain, repeat the process.

Oil and grease stains. Try wiping off with paint thinner or benzine, but don't scrub so hard that you force the oil further into the wood. If this doesn't produce results, saturate cotton with hydrogen peroxide and place over the stains. Then soak another piece of cotton in household ammonia and spread on top. When the cotton dries out, repeat the process until the stains disappear.

Chewing gum stuck to floor. Chill the gum by holding an ice cube on it until brittle. Then scrape off with a dull knife. Remove residue with trichloroethylene or another nontoxic cleaning fluid.

Candle wax. Scrape off with a dull knife and scrub with trichloroethylene.

Heel and caster marks. Rub with white liquid wax of the type used to clean and polish kitchen appliances. The alternative is to rub with fine steel wool and a waterless floor cleaner.

Mildew. Despite the basic rule against applying water or a water-base substance to wood floors, mildew is so troublesome once it gets started that it's best to nip it in the bud as soon as you can, and the surest way to do this is to scrub it off with a rag well wrung out in a strong solution of chlorine bleach. Then rinse with a rag well wrung out in clear water, and dry thoroughly.

Paint. If you see the spatters in time, wipe them up immediately with the appropriate solvent (water for latex paint; mineral spirits for alkyd and oil-base paints; alcohol for shellac and alcohol-base stain killers; lacquer thinner for lacquer). If the paint dries before you find it, scrape up as much as you can with a razor blade or sharp knife, and leave well enough alone. The only way you can get the pores of the wood clean is to sand the area, and this is likely to leave an even more noticeable spot.

Burns. Unless burns are very superficial, there is no possible way of removing the marks without cutting into the wood. This means you end up with a cavity that should be filled if very deep—and there's no filler that will match the wood exactly. In short, burns are anathema to anyone who takes great pride in his floors.

To remove the char, start with a knife with a small blade rounded at the end and slowly scrape out the blackened wood down to the brown. Wipe out as much of the deep brown color as possible with benzine or paint thinner. Then, to lighten the brown further, apply bleach. When you finally arrive at a color you consider acceptable, sand the hole—just the hole—and apply whatever finish you have used on the surrounding floor.

If you prefer filling the hole—regardless of the results—buy colored plastic wood to match the floor. Theoretically, this is available wherever ordinary plastic wood is sold, but I have never found a store that carried all the colors made. So you may have to get the proprietor to order it for you. The alternative is to mix ordinary plastic wood—which dries to a nasty whitish-yellow—with a little alcohol stain the color of the floor. Then—whichever material you use—pack it into the sanded hole and smooth the top. Sand when dry, and apply floor finish.

Holes. Fill pits with colored plastic wood as above.

Actual holes in floors are rare, but you run into them occasionally in old houses where radiator and water pipes have been taken out. Since there is no subfloor underneath, the only way to fill the hole is to draw a neat square or rectangle around it and cut it back to the lines with a rasp. Bevel the edges so the hole is smaller at the bottom than at the top.

Out of a scrap of wood matching the finish flooring, cut a piece of the same size and bevel the edges in the same way. Test it in the hole until it fits exactly, with the top flush with the floor. Then coat the beveled edges with white glue and weight the plug down in the hole for about six hours. It can then be sanded as necessary and finished.

Splinters. Glue these down with white glue as soon as you notice them—before the wood is ripped away any further or is broken into fragments and disappears.

7/ WOOD FLOORING-PLANK

Of all wood floors, plank floors are the handsomest.

Living in the lower Connecticut River Valley, I am figuratively surrounded by some of the most beautiful pegged and unpegged plank floors man has ever created. Made primarily of eastern white pine, their gentle brown color and quietly glowing patina have the same exciting influence on the spirit as a neat lawn on a spring morning. No artist has ever created such beauty or feeling in a monochrome.

Adding to the appeal of the floors is the incredible width of some of the planks used. Twelve-inchers are nothing. Fifteen-, 18-, 21-, and 24-inchers are commonplace. A board 2 feet wide? You're kidding. It's gotta be plywood. But if you were to pry it up from the joists, you'd find it is not only solid wood but probably 1¼ or 1½ inches thick. Skimpy boards weren't tolerated by our forefathers. They wanted them thick enough so they could omit a subfloor. If the boards were also thick enough to last for ten or twelve generations, so much the better.

I get much the same feeling from the plank floors in my cousin Mike's house in Mississippi. True, his floors are only fifteen years old and the boards themselves are toothpicks in comparison with some of those in Connecticut, but they have an exciting quality you'll never see in a strip floor or, for that matter, in many plank floors. They're made of black walnut. Sacrilege! you say? Well, look at it this way. If you enjoy feasting your eyes on one of the world's most richly beautiful woods, why not use it in a great expanse that is within your line of vision throughout a goodly part of every day? Gorgeous woods weren't put on the earth to be coddled and given hands-off admiration like a *Mona Lisa*. They're working materials first and foremost, and if you can afford them, a floor is a first-rate place to use them.

Here's the secret of a plank floor's beauty: It shows off the wood better than any other kind of floor. You don't see just a sliver of board; you see the whole board in all its glory. And if you're moved to study a floor, you find that each board is a distinct entity to be enjoyed as much for itself as for the contribution it makes to the overall floor.

The other features of a plank floor are of secondary value, though not to be dismissed. The pegs that theoretically keep the boards fastened down are more nostalgic than anything else. But they break the flow of the floor and give it interest, too. I have noticed more than once how women seeing a plank floor for the first time often single out the pegs for comment.

The V-joints that are commonly—but not invariably—found between the planks give the floor texture, and at the same time—like the frame around a picture—they focus attention on the individual boards.

Woods for plank floors.

Talk of floors made of black walnut and 24-inch white pine is not meant to suggest that all plank floors should be made of similarly exotic materials. The vast majority, in fact, are made of red or white oak. I don't think oak is plebeian, but on the other hand, considering how much it's used in strip flooring, you can't call it exotic, or even unusual.

What I am trying to suggest is that, if you're desirous enough of beauty to choose a plank floor over a strip floor, you might consider going one step further and having it made of a particularly lovely wood you don't encounter everywhere: teak, for instance, or wormy chestnut, or red cedar, or even mahogany.

Unfortunately, few flooring manufacturers and dealers offer anything other than oak, but there are some with a bit more imagination. One of the largest manufacturers in Tennessee, for example, offers white and red oak (both either quarter-sawed or plain-sawed), cherry, black walnut, and maple. An importing firm in Philadelphia sells teak and something called Asian ironwood.

And if you can't find what you want ready-made, you can usually have it produced to order through or by a good lumber dealer.

Sizes and types.

The standard floor plank is made of solid wood ¾ inch thick, comes in random widths, is tongued and grooved along the edges and at the ends, has chamfered edges to form a V-joint between adjoining boards, and is unfinished. This is the type you will get if you call up a lumberyard and say, "I want material for putting down a random plank floor in my fifteen- by eighteen-foot family room." But you can also buy:

Planks in other thicknesses down to 5/16 inch.

Planks in a single width, or in two widths so you can lay them in an alternate wide and narrow arrangement.

Planks with square edges and ends (without tongue-and-groove joints or chamfered edges).

Prefinished planks.

And laminated planks (made of two thicknesses of wood bonded together) which are recommended especially for installation in damp locations.

If the planks are ordered from a large manufacturer, they are shipped in bundles containing assorted lengths ranging from about 2 to 8 feet and widths from 3 to 8 inches. On the other hand, planks produced to order by a local sawmill or lumberyard come in much greater lengths and are also available—if you ask for them—in greater widths.

Since this lack of uniformity makes for confusion, the best thing to do when buying plank flooring is to put yourself in the hands of a local lumber dealer. For example, several years ago when I decided to lay a random-width plank floor in a hall, I told my usual lumber dealer simply that I wanted standard-thickness planks without chamfered edges in 6-, 8-, and 10-inch widths. For reasons I still don't understand (because they could have ordered the material from a number of national flooring manufacturers), they replied that all they could offer were planks with chamfered edges. So I called another smaller yard, which I knew sawed lumber to order. "No problem," they answered. "But we'll have to run the boards off for you, so you won't get them for a couple of days. Furthermore, they won't be tongued and grooved. Boards that aren't chamfered have square edges. The only boards we carry in stock for immediate delivery are tongued-and-grooved and chamfered."

Estimating your needs.

Like most other lumber products, plank flooring is sold by the board foot. Rather than going to the trouble of figuring out how many board feet of planks you need in each width desired, just tell the dealer that you want enough material to cover so many square feet. To determine the square footage required, measure the width and length of the room and multiply the figures. Then measure the square footage of closets and alcoves and add the total to the figure for the room. When placing your order, be sure to specify that you need so many *square feet* of flooring.

At the same time, give your order for the wood plugs—sawed-off pegs—you need. This averages out to roughly 1½ pegs per square foot in a tongue-and-groove floor, 2½ in a square-edged floor. The plugs are made of the same wood as the planks (although you can use another wood for special effect) and are ¾ inch in diameter.

Laying a plank floor.

Whether you lay planks over a subfloor, an old finish floor, or sleepers, the preparations are like those for a strip floor. The planks are also arranged in the same way—usually lengthwise of the room and with ½-inch open expansion joints around the perimeter.

Before taking hammer in hand, figure out the sequence in which you lay the different widths of plank across the room. Let's say you'll start with a 6-incher, follow with a 10-incher, and then use an 8-incher, and repeat this arrangement all the way. Or you might start with an 8 and follow with a 6 and 10; then use an 8, 10, and 6; then go to a 6, 8, and 10. From the standpoint of appearance, the actual sequence is unimportant: You can repeat the same sequence ad infinitum, skip back and forth, or clump up a couple of boards of the same width. The important thing is to plan the layout of the planks so that you won't have to cut one of the wide planks into narrow strips when you put down the last row.

The end-to-end arrangement of the planks is effected as you lay the floor. Some of the planks may be long enough to extend all the way across the room. Other rows must be made up of several short planks. In the latter case, try to place the shortest pieces in the center of the room, and if the planks are not tongued and grooved at the ends, take pains to cut them square so the ends butt tight together. End joints in adjacent rows of planks must be staggered at least 6 inches.

Tongue-and-groove flooring is fastened down, like strip flooring, with screw nails driven diagonally through the tongues. Then, to make sure the boards will not buckle upward at the center, drive 1½-inch flathead steel screws through the faces. (This can be done after all the boards are nailed down.) At the ends of each board, place one screw in each corner 1 inch back from the edges and about 1½ inches from the ends. These are not actually needed, but they make the boards look as if they are pegged down like those in very old houses. The other screws can be placed at random—some in the centers of the boards, some to either side of the center lines. Some can even be placed along the edges—again to give the effect of real pegs. Space the screws 2 to 3 feet apart along the length of the boards. In wide boards, which are especially prone to buckling, the spacing should be closer than in narrow boards.

Square-edged flooring is held down entirely by screws; consequently, after you have shoved each board tightly into position, you should install the screws at the ends and at least some of those in the middle. Here, again, there is no definite screwing pattern that must be followed; however, to keep the boards from cupping upward at the edges, you need to place screws at fairly frequent intervals along both edges, and you also need screws down the centers of the boards to prevent buckling. For example, starting from the two screws at one end of a board, you might place a single screw in the center of the board 2 feet from the end, then place a pair of screws near the edges 2 feet further along, then place a single screw in the center 2 feet from these, and so on. Just remember that the screws are the only things holding the boards in place, so don't stint on the number. On the other hand, don't use so many that the floor looks as if it had chicken pox.

Of course, the screws used in a plank floor are not exposed. They're hidden, on the contrary, by the plugs which simulate the pegs in old floors. This means you must drill holes for the screws and also drill holes for the plugs. For such a tedious job, you need an electric drill. Then, to take still more of the tedium out of the work, you should buy a special type of drill bit called a countersink-counterbore. In

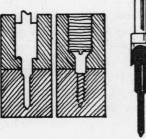

one operation, this drills a pilot hole for the screw shank, a slightly larger, conical hole for the screwhead, and a ¾-inch hole for the wood plug.

After drilling holes through each plank, drive the screws down tight into the subfloor or sleepers. A ratchet screwdriver saves time and energy. Then set the plugs over the screws. Most of them will fit tight. If there are any that don't, put a little white glue on the sides and bottom.

Sanding and finishing plank floors.

Because of the width of floor planks and because they usually have chamfered edges any slight unevenness in a completed floor is not obvious to the eye or feet. It follows that if the planks you buy were well sanded at the mill and have not been damaged during installation, there is no necessity for renting machines to sand the entire floor. All you have to do is sand out any imperfections in the planks and sand down the plugs flush with the surrounding wood. This you can do by hand. If the plugs project too much for easy sanding, cut them down with a very sharp block plane.

If a new or old floor requires overall sanding, follow the procedure for sanding strip floors.

Finish the floor with a penetrating sealer.

Laying plank flooring in adhesive.

Planks that are suitable for gluing are ⅜-inch-thick, laminated (plywoodlike) materials, which are preplugged and finished before they leave the factory. They are sold in rather narrow widths (up to 7 inches) and short lengths (2 to 5 feet) for easy handling. Installation can be made over concrete, plywood, old wood floors, and resilient tile.

The room in which the floor is to be laid should be warmed to 70°, and the flooring and adhesive should be stored there for at least twenty-four hours before installation.

If applying the planks on concrete, the concrete must be dry, level, smooth, clean, and free of grease and oil. Grind down rough and high spots with a terrazzo grinder. Fill low spots with latex cement troweled to a feather edge.

Any plywood subfloor is satisfactory as long as it is sound and level. Small voids in the surface ply need not be filled. Adhesive-laid floor planks must not, however, be installed directly on a board subfloor or finish floor. After nailing the boards securely to prevent squeaking, cover them with ⅜-inch plywood or particleboard.

Resilient tiles can be planked over only if they are well bonded to the subfloor. Remove all wax and roughen the tiles with sandpaper. Reglue loose tiles (but if there are many of these, it's preferable to take them all up and scrape off all traces of the old adhesive).

The best procedure in laying the planks is to strike a chalk line across the floor 2 feet in from the door through which you enter the room. Lay the planks from the line back to the far wall, and then from the line forward to the door wall. In other words, the 2-foot area in front of the line serves as work space when you start laying backward from the line. From then on, the newly laid planks serve as your working platform.

Use the adhesive specified by the flooring manufacturer and apply it with a notched trowel recommended by the manufacturer. Since the adhesive dries slowly and the planks go down quickly, you should be able to lay 100 square feet of planks within the adhesive's two- to three-hour drying period. So, starting at the chalk line, apply the adhesive in a 100-square-foot strip, and so on until the entire area behind the line is planked. Finally, complete the area in front of the chalk line.

After spreading adhesive evenly on the subfloor, let it set for about forty minutes or longer until it is tacky enough to grip the planks. Lay the first row of planks with the tongue next to the chalk line. Press them down firmly so they don't skid when you walk on them or push against them. When laying subsequent planks, engage the tongue-and-groove joint at the end of the plank, then slide the plank toward you until the side tongue is locked in the groove of the plank already laid. Try not to get the adhesive on your fingers: Wipe your hands frequently with paint thinner. If you can't make a tight joint by hand, hold a block of wood against the edge of the new plank and tap the plank into place with a hammer. After you've laid about 25 square feet of planks, roll them with a 150-pound flooring roller from a rental agency or the flooring dealer.

Provide ½-inch expansion joints between the edges of the flooring and walls. To help fasten the planks down and also to improve the appearance of the floor, stagger all end joints between adjacent rows at least 6 inches. As with nailed plank flooring and strip flooring, try to use long planks at the ends of the room, short planks in the center.

Wipe up adhesive that gets on the surface of the flooring as your work progresses. Then, at the end of the day, inspect the entire floor and remove remaining smears with paint thinner. Finally, when the entire floor is down, give it a coat of solvent-base paste or buffable liquid wax, and buff well.

Maintaining and repairing a plank floor.

Everything in the preceding chapter about maintaining and repairing a strip floor applies to a plank floor. You may also run up against several problems peculiar to plank floors.

When Elizabeth and I bought our present house, one of the first things we noticed (we couldn't help noticing) was the poor state of the living room floor. Because the house had been unoccupied and the crawl space under the living room was not well ventilated, several of the wide boards had cupped so badly that they created sharp ridges about ½ inch above the floor level.

The normal treatment for floor planks that are slightly cupped is to drive screws through them, close to the up-curving edges, and tighten the screws gradually—over a period of several weeks—until the boards are more or less flat. But when the planks are as badly cupped as ours were, screws alone rarely work. The first thing you must do is to saw through the joint between the raised planks so they are no longer wedged into a ridge by the tongue-and-groove edges. Use a power-driven saber saw or circular saw and run it the entire length of the raised joint. Then try to flatten the boards by driving screws through them into the subfloor and, if possible, the joists.

If a sharp ridge persists, the only thing you can do is to saw out the bulge or to remove the planks on either side of the bulge. These obviously are drastic measures, and you need a circular saw for them. Adjust the blade height so it will cut through the finish flooring but won't penetrate deep into the subfloor. Then turn on the power, tilt the saw down to the floor and let the saw cut gradually through the board, then move the saw along or across the boards. Remove the cutout pieces and nail new flooring into their places.

Sawing out the bulge is easier than removing entire sections of plank. There's no possible way of making the repair attractive; therefore it must be hidden under a rug. In fact, when I removed the worst ridge in our living room, I simply filled the gap with a pine board cut roughly to shape. Now that it's covered by a rug cushion and carpet, you would never know the patch was there.

The only reason for sawing out entire planks on either side of the bulge is to fool people into thinking the floor is as good as new, and for this reason you must take pains to put in a patch that gives no suggestion of being a patch. That's not an assignment to be undertaken lightly.

Wide gaps between planks are a very common affliction of old floors, especially those of pine. In an earlier house, I tried every conceivable way of filling the gaps, and only one of them worked. Ordinary—and some not-so-ordinary—filler materials such as plastic wood, spackle, sawdust soaked in glue, rope and oakum, and conventional caulking compounds were complete failures. They simply couldn't withstand the flexing, contraction, and expansion to which they were subjected.

The only way to fill gaping joints in a plank floor is to shape strips of wood to fit, and fasten them in with glue.

8/ WOOD FLOORING – PARQUET

Parquetry is mosaic work done with wood. A parquet floor is therefore a wood floor composed of small pieces of wood fitted together in a pattern that is repeated over and over again.

There is no other kind of floor—not even a ceramic tile floor—that is more elaborate. That's why parquetry was found only in mansions and handsome public buildings from the time it was first used in Europe in the fourteenth century until only thirty or forty years ago. There was a time long ago, in fact, when builders and architects vied to see who could come up with the most unusual, most beautiful, and sometimes most bizarre parquet floors.

All that is changed today. If you're so minded, of course, you can still build ornate parquet floors—at a cost of about $8 a square foot and up. But most parquet is now relatively simple and geometrical—and the cost has dropped accordingly to as little as $1 a square foot. With this change, parquet has soared in popularity.

There are several reasons why this has happened. One—probably the key—reason is that so many houses are now built on concrete slabs, and wood blocks, unlike strips and planks, are ideally suited to installation on slabs because they can be laid in adhesive.

Another reason is that homeowners today are pattern-minded. They like materials that produce interesting and attractive designs.

And the third reason is that the method of laying parquet floors has been greatly simplified. Originally these floors were put down slowly and painstakingly, small piece by small piece. Today, as a rule, the small pieces are assembled at the factory into tilelike blocks that cover as much as 2½ square feet of floor at once. (To be technical about it, this kind of flooring is called block flooring rather than parquet flooring, but since the effect is the same as that of parquet, the distinction is rarely drawn.)

I've never seen a house in which parquet flooring is used throughout the living and sleeping areas. This doesn't, of course, mean that such houses don't exist. But the cost mitigates against the idea. More important, parquet is too lovely to be used indiscriminately. It should, rather, be used in individual rooms to impart a special air, a special feeling of distinction. For this reason, it's ideal for front halls, living rooms, dining rooms, libraries. It's also good in rather formal family rooms. But in the average family room that gets a lot of rough-and-tumble wear, it's simply too difficult to keep in proper condition. And for most bedrooms, even the simplest design is too formal.

Woods and patterns.

Although I haven't explored the idea in depth, I venture to say that no other single building or decorating material can be produced in so many ways as parquet flooring. Not only are there thousands of tree species whose wood might be used, but there is no limit to the number of designs that man can create. If this makes your imagination soar, let it.

But even if you prefer to stick to the ready-made parquet that's on the market, you'll probably have trouble deciding which wood and which pattern to choose. There are so many.

The basic woods are oak, maple, beech, birch, black walnut, teak, mahogany, pecan, ash, cherry, cedar, ebonized wood made by dying white maple or holly, and about a half dozen imported woods that have been given cute registered names that mean nothing to the botanist.

Patterns start with a standard design, in which narrow strips of wood are laid side by side to form squares that are placed in checkerboard fashion. They include the well-known herringbone. And they go on to a delightful array of squares, diamonds, basket-weaves, rhomboids—you name it. One firm alone advertises over a hundred different designs for immediate delivery.

A unique product made by ARCO Chemi-

cal Company* is an 11⅜- x 11⅜-inch tile made of particleboard impregnated with acrylic and colored blue, green, or orange-red as well as several shades of brown. Strictly speaking, this has a closer resemblance to resilient tile than to parquet, but I include it in this chapter because it performs like wood.

Sizes and Types.

If there's anything standardized about parquet flooring, I have yet to find what it is.

Assembled blocks range from roughly 6 to 39 inches square, single strips from 1 x 6 inches to 3¼ x 16 inches.

Although the majority of blocks are thin, thicknesses range from 5/16 inch up to 33/32 inch.

Some units are tongued and grooved, others are square-edged.

Some blocks have chamfered edges to make them stand out in a floor as individual entities; others do not.

Blocks are either unfinished or prefinished.

Prefinished blocks are generally held together by a fabric backing glued to the underside. Some unfinished blocks also have fabric backing, but most are surface-covered with paper that must be moistened and pulled off after the blocks are laid.

---

* Centre Square, 1500 Market Street, Philadelphia, Pa. 19101.

Estimating your needs.

Since parquet flooring is sold by the square foot, all you have to do to determine how much you need is to figure the exact square footage of the floor and add 5 percent for waste in cutting. Divide the total by the number of square feet of flooring contained in a carton (this varies).

Order the necessary adhesive from the flooring dealer, too. Not that his is any better than anyone else's; it just saves time and trouble to buy everything from one source. You'll need one gallon to cover 40 to 60 square feet, depending on the brand.

Laying parquet flooring.

Although parquet flooring can be put down with nails driven diagonally through the tongues, it is usually glued. Handled thus, it can be applied to just about any base that is sound, level, smooth, and clean.

Every concrete slab must be allowed to cure for at least thirty days before flooring is laid. You should then check it for alkilinity and moisture by having a druggist mix 3 percent phenolpthalein with 97 percent grain alcohol. Scatter a few drops on the slab here and there. If they turn red, the flooring installation should be delayed until further tests produce no color change.

The slab should then be leveled and smoothed. Use a terrazzo grinder to take down high spots and rough spots. Trowel latex cement into low spots. Remove paint, wax, and other foreign coatings.

If the slab is suspended, no further preparation is called for. But slabs laid on the ground must be dampproofed to protect the flooring against moisture, which might later penetrate the concrete, loosen the adhesive and rot the wood.

A slab on grade may be considered safely dampproofed if you know beyond the shadow of a doubt that it was poured on a polyethylene moisture barrier. If you're not positive about this, however, the slab must be topped with a special dampproofing membrane. This can be constructed in either of two ways:

1) Prime the concrete with a primer recommended by the flooring manufacturer. When dry, coat it with the flooring adhesive. This is applied with the smooth edge of a flooring trowel. After this has dried slightly (follow the flooring maker's directions), roll in it a 2-mil polyethylene sheet. Lap the edges of adjoining strips 2 to 4 inches. Smooth the film as evenly as possible into the adhesive, but don't worry about scattered small blisters. Then proceed with the flooring installation.

2) Prime the concrete with the flooring manufacturer's primer. Over this apply an even coat of hot asphalt or special mastic recommended by the manufacturer. Immediately embed a layer of 15-pound asphalt-saturated building felt. Butt the edges and press out blisters. Coat the felt with hot asphalt or mastic and embed a second layer of felt. Start with a half-width strip so the seams in the first layer will be covered. Then install the flooring.

If you're a gambler, below-grade slabs can be dampproofed in the same way. But in view of the cost of a parquet floor, it's better to call in a professional flooring contractor and abide by his recommendations.

Parquet can be laid on a plywood subfloor provided it is at least ¾ inch thick. The plywood must be securely fastened to the joists in accordance with the nailing schedule in Chapter 2. Cross blocks that are 2 x 4 inches must be installed between joists under the unsupported edges of the plywood.

On a subfloor made of boards or on an old wood floor, apply an underlayment of ¼-inch hardboard or plywood. See directions for installing underlayments for resilient floors in Chapter 10.

Resilient flooring materials should always be removed, despite the fact that some parquet manufacturers approve making installations over them if they are tightly cemented. Scrape, sand, or wash off the old adhesive on the subfloor. If the subfloor is surfaced with plywood or hardboard, the parquet can be laid directly on it, but if the subfloor is made of boards, apply a ¼-inch underlayment.

In all cases where the subfloor is made of plywood or hardboard, it is unnecessary to apply a primer before laying parquet.

After the parquet is delivered, give it at least twenty-four hours to become acclimated to the room in which it will be used. Turn the heat up to 60° or higher.

Parquet flooring is generally laid in rows perpendicular to the walls, but it may be laid diagonally. The first step, in either case, is to divide the room into quarters by striking chalk lines from end to end and side to side. The lines should cross at the exact center of the room.

If using a square pattern—that is, installing the rows of blocks perpendicular to the walls—you should then adjust the chalk lines so that blocks around the edges of the room will be approximately equal in width. This process is described in Chapter 10. Either one of the intersecting lines may serve as the starting line from which the blocks are laid.

If you're using a diagonal pattern, measure 4 feet from the center of the room along each line toward the four walls, and mark points a, b, c and d (see drawing). Tie a string around a nail and tie a pencil to the other end of the string exactly 4 feet from the nail. Have an assistant hold the nail at point a and then at point b while you draw intersecting arcs between these points at A. Then draw intersecting arcs between c and d to establish point B. Snap a chalk line across A and B from wall to wall. Start laying blocks from this line until the half of the room more distant from the entrance is covered. Then complete the second half.

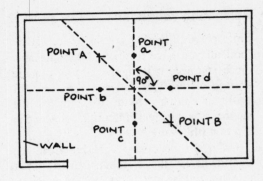

Whether you use a square or diagonal pattern, lay the first block at the center of the room flush with the starting line. Lay the next two blocks on either side of the first and on the same side of the starting line. Build up and out from these (as in the drawing) in a constantly enlarging pyramid until the walls are reached. Then lay the floor away from the starting line in the opposite direction but in the same way.

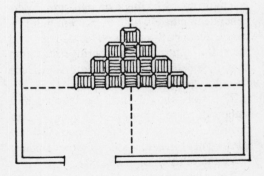

Because of the many variations in the construction of parquet blocks, there is no single set of rules for their installation. Follow the instructions that come with them. The directions below serve only as a general guideline.

1) Apply adhesive to the subfloor just ahead of your work. Put down no more than you can cover in the next ten to fifteen minutes. Spread the adhesive with the notched edge of your trowel. Hold the trowel at an angle of about 45° and scrape it across the subfloor so that little or no adhesive is left in the valleys between the ridges. The ridges, however, should be full and thick. To make sure they are, clean out the notches in the trowel frequently.

Check occasionally whether you are applying enough adhesive by lifting out a block you have pressed down and examining the underside. It should be well covered.

2) Lay the blocks close together, but don't hammer them tight. Fit them with your fingers only.

When placing a block, drop it as closely as possible into position. Sliding it makes the adhesive pile up on the leading edge, thus preventing a good fit.

If any adhesive squeezes out on the surface of the blocks, wipe it off with paint thinner.

3) Press the blocks firmly into the adhesive once they are positioned. Hand pressure is usually sufficient, but you may use a rubber mallet or a smooth block of wood under a hammer.

Rolling areas you covered thirty minutes previously is a good practice, though not essential. Use a 150-pound flooring roller.

4) If the block faces are covered with paper, wipe the paper with a damp sponge before laying each block. For speediest removal of the paper, dissolve about 3 ounces of a liquid detergent such as Fantastic in 5 quarts of water, and dampen—don't soak—the sponge in this. Lay the block at once.

After about five blocks have been laid, dampen the paper again; then remove it by peeling it back over itself—don't pull straight up—to avoid loosening the blocks. Adjust any of the segments that may have slipped out of line, and press or tap them down.

5) Like other wood flooring, parquet requires expansion joints at the wall lines. Unless the manufacturer's directions stipulate otherwise, allow 1/16 inch space for each foot of floor and divide the space equally at each wall line. That is, in a 12- x 20-foot room, you should allow a total of ¾ inch expansion space at the side walls and 1¼ inch at the end walls.

If a block next to a wall must be trimmed, spread a piece of plastic kitchen wrap over the adhesive at the foot of the wall. Place a block (marked A in sketch) squarely over the block in the preceding row. Place a second block (marked B in sketch) on top of A and shove it against the wall. Draw a line on A along the edge of B. Cut block A on this line with a fine-toothed saw. Be sure to hold the block firmly on a flat surface so the saw won't loosen the segments. Then remove the plastic wrap and press the block into position.

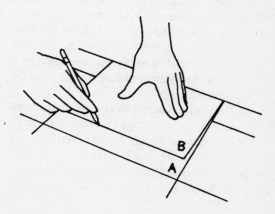

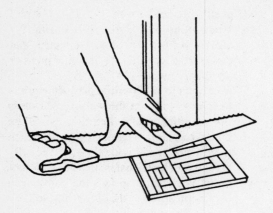

6) Where a block abuts the trim or jamb in a doorway, it must either be cut to fit around the trim or slipped under it. The latter method is preferred since it assures a neater fit without dust-collecting cracks. To accomplish the fit, put a piece of a block on the subfloor next to the trim, lay a saw on top, and cut off the bottom of the trim flush with the top of the block. Don't spread adhesive under the trim. Simply slide in the block.

7) If parquet is laid over an existing floor so the surface is higher than the floor in an adjoining room, the exposed edge at the juncture of the rooms must be protected with a strip of wood butted to the edge and nailed to the lower floor. Some parquet manufacturers provide special thresholds that match the parquet. If these are unavailable, you should buy a length of strip flooring matching the parquet and trim it to fit. Bevel the edge facing away from the parquet.

8) Similarly, parquet should not be allowed to extend to the edge of a step down from the room in which it's laid. Nail a board along the edge and butt the parquet to this. You may find that the flooring manufacturer has step nosings made for the purpose.

9) Do not step directly on newly laid parquet for about six hours. Cover it with plywood.

Laying a herringbone floor.

This popular type of floor is laid like other parquet floors. The layout of the strips is the only tricky part.

Establish the center of the floor with intersecting chalk lines. Then draw a line at a 45° angle through the center point.

Place the first herringbone strip along this line with the corner at the center of the floor. Butt the end of the second strip to the side of the first to form a V. Set the third strip into the V parallel to and against the side of the first. Then set the fourth strip into the newly formed V parallel to the second strip. And so on until you reach the wall.

From here, work forward to the center line, fitting strips into the notches at the sides of the first two rows.

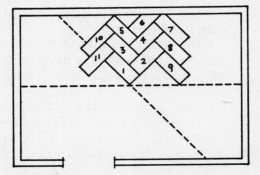

Sanding and finishing parquet flooring.

Prefinished flooring, of course, requires little work. Let the adhesive set overnight. Nail the baseboards and/or shoe moldings to the walls. Remove adhesive from the floor with paint thinner. And apply a coat of paste wax or solvent-base buffable liquid wax.

Unfinished parquet must not be touched for twenty-four—preferably forty-eight—hours after the last piece has been glued down. You should then level and smooth it with floor sanders. Despite the fact that the wood grain runs in two or more directions, the work isn't a great deal different from the sanding of a strip floor. You can do it with a drum sander and an edger with a disk at least 14 inches across, or you can use a 14-inch edger alone.

If you use a drum sander (it saves a little time), use it with coarse and medium sandpaper on the center of the floor. Sand at 45° to the pattern. Then use coarse and medium paper on the edger to get close to the baseboards. Then do the entire floor with the edger, using fine paper.

The floor is now ready for finishing. Some perfectionists, however, go one step further and polish the bare wood with fine steel wool under a rotary buffer.

Finish the floor with a stain (optional) and penetrating sealer.

Maintaining and repairing a parquet floor.

Follow the directions for taking care of a strip wood floor (Chapter 6).

9/ WOOD FLOORING-BLOCK

If you have ever walked through factories and warehouses, you have seen floors made of brick-size wood blocks. But you wouldn't consider them for your home? Don't be too sure of that. They're ideal for a workshop that gets heavy use. They would be excellent in a basement that is used for ping-pong and young children's rough-and-tumble play. And believe it or not, they would be beautiful in a family room or certain kinds of kitchen.

No, I am not out of my mind.

Wood blocks, despite their stolid solidity, have a surprising resilience—and that spells comfort underfoot. They're as skidproof as concrete, and not half so lethal if someone falls. They're excellent insulators against cold seeping up through concrete subfloors. They can take all the punishment you give them and show no more signs of wear than bricks. And as for being handsome enough for a family room—I am not comparing them with marble, but if you give them the right finish, you might well mistake them for a slightly rustic parquet.

Cut from the end grain of yellow pine or oak lumber, the blocks range in size from approximately 3 x 6 inches to 4 x 8 inches. Thicknesses run from 1½ to 4 inches, but for residential use the 1½- or 2-inch thickness is ample. Because of their appearance, the best blocks for homes are simple rectangles (others have vertical grooves or beveled corners).

Installation is made over a poured concrete or solid masonry subfloor. As long as this is level, clean, and free of ridges, it doesn't make much difference how rough it is. Prime the floor with the material specified by the block manufacturer, and when this is dry, trowel on a layer of the companion adhesive. Lay the blocks in the adhesive close together, with the grain vertical. Use a running bond as for bricks (see Chapter 14). Provide 1-inch open expansion joints between the floor edges and walls.

Wait at least two days before finishing the floor. First you must sand it with a drum sander using no. 3½ paper and then with a disk sander using similar paper. Vacuum thoroughly. Mop on a coat of clear sealer recommended by the block manufacturer; let dry for four hours and seal again. When the second coat has dried, mop or spray on a coat of the manufacturer's urethane finish and let dry for twelve hours.

The joints are then filled by squeegeeing into them a special filler from the block manufacturer. Finally, two additional coats of urethane are applied. For a pluperfect finish, go over the first coat of urethane with a disk sander using fine paper before applying the second coat. Wait two days before walking on the floor.

Because of the elaborate finishing process, a wood block floor needs little maintenance. Regular vacuuming and occasional waxing with a solvent-base paste or buffable liquid wax will do the trick.

If a block is damaged, pry it out and replace with a new unit set in the proper adhesive or in silicone bathtub caulking. Complete sanding and refinishing are called for only when the entire floor becomes badly worn.

Stains are treated like those on a strip wood floor (Chapter 6).

10/ **RESILIENT TILES**

Most homeowners get into floor laying by installing resilient tiles in a kitchen or bathroom. It's a splendid introduction to a not-terribly-difficult craft. With the possible exception of bricks that are laid without mortar, resilient tiles are the easiest-to-lay flooring material, and they are inexpensive enough not to punish you for mistakes you make.

I remember that when I laid my first asphalt-tile floor (in the kitchen, because it was the only thing we could afford), I ruined a half dozen or more tiles during cutting, and there were several places where the installed tiles were uneven. But by the time the floor was completed, I felt capable of tackling anything in the flooring line. And when we finally sold the house fifteen years later, it tickled my soul to see that the floor was still attractive and in excellent shape.

Such are some of the benefits of resilient tile flooring. Others are more important.

To begin with, resilient flooring (I include sheet materials along with tiles) offers more color and design than any other permanent flooring material except ceramic tile. This doesn't mean that all of it is beautiful: When

man sets out to develop designs with great market appeal, various weird ones are bound to result. I have known years, in fact, when I thought almost all the designs were atrocious. But by and large, the selection is tremendous. You can find a pattern suitable for every room in the house.

Despite its beauty, resilient flooring is amazingly tough, wear-resistant, and durable. Many floors in heavily trafficked commercial buildings are still in good condition after thirty years. But here again I must be careful because there are also floors that have worn out all too quickly. I had one of these and I must admit it has turned me forever against that particular type of flooring. But if you think one unfortunate experience would make me give up resilient flooring in general, you couldn't be more mistaken. It's *good* stuff.

The easy cleanability and general maintenance of resilient flooring are outstanding. Admittedly, many housewives will dispute this point. I have had my share of complaints and questions from them. But in virtually every case either they or the men who installed

70

the flooring were at fault—not the flooring itself. Put down carefully and maintained according to the rules, resilient materials are about as easy to care for as any flooring made.

As its name suggests, resilient flooring is softer underfoot than many other materials, and being softer, it is also quieter. Unfortunately, it also dents more readily, and in some cases, the dents never disappear.

As for resilient tiles versus resilient sheets, both have their good points. The tiles, as I said earlier, are easier to lay—much more so than sheets. And they can be used in an infinite array of geometric patterns. On the other hand, since sheet flooring is almost seamless, it is slightly more waterproof and easier to clean. And you can create inlaid designs ad infinitum.

Types of resilient tile.

Solid vinyl, though not common, is the best. Laid on grade, below grade, or on suspended subfloors, it is very durable and shows little signs of wear. It is very resistant to damage by grease and has good resistance to denting and staining. It is more resilient and quieter than all other tiles except cork (though it is nowhere nearly as resilient and quiet as sheet vinyl). It is also reasonably easy to maintain. Its one weakness is its poor resistance to cigarette burns. Cost starts at approximately 75 cents a square foot.

Vinyl-asbestos tile is the most popular of the resilient tiles because you can get if for as little as 35 cents a square foot. Like vinyl tile, it can be installed at any level, but in most respects it ranks a notch below vinyl. Its strong point (next to low cost) is quite high resistance to burns.

I must confess, however, that vinyl-asbestos is the one resilient tile I'll no longer have in my house. That is quite a switch,

because when it was first introduced, I thought it was in a class by itself. But when I began to notice that the best available vinyl-asbestos tiles were wearing out only four years after they were installed in our present kitchen, I had a change of heart. Probably we just got a bad batch (I know it wasn't an improper installation). But experiences like that don't sit well.

Asphalt tile seems to be on the way out. This is a pity because, although it's available in only a limited and dull array of patterns and colors, there isn't anything better for installation on or below grade, and it lasts and lasts and lasts. But it has quite a few strikes against it otherwise. It's hard to keep in mint condition; it stains and dents pretty badly; it shows burn marks; it is no more resilient or quieter than ceramic tile; and it cracks readily. Cost: about 30 cents a square foot.

Rubber tiles are today made by only two firms. These tiles never were terribly popular, partly because the colors are not very bright, and partly because of their cost (today, about $1.20 a square foot). In addition, they have only fair resistance to abrasion and grease. But they are as comfortable and quiet as you can ask for, and they rarely show dents.

Cork tile is the most resilient of all resilient floorings, and is therefore especially desirable where you want a floor that is very kind to feet and knees and/or want to muffle the sound of footsteps. But it does not wear as well as other tiles, it bleaches out in direct sunlight, and it stains badly. To increase resistance to bleaching and staining, most tiles are now treated at the factory with a plastic finish. They are available in light (nearly natural), medium, and dark tones of brown. Given good protection, they are as beautiful as wood.

The most popular thicknesses for flooring are ⅛ and ⅝ inch. The former costs approximately $1.25 a square foot.

Estimating your needs.

Almost all resilient tiles measure 12 x 12 inches. Only a few of the once-standard 9x9s are made, and most of those are asphalt. A tiny handful of tiles are more than a foot square. Thicknesses range from about 1/16 to ⅛ inch. For long life, the thicker the better.

To determine how many tiles you need for a job, measure the width and length of the room and multiply the figures to find the square foot area. Find the square foot area of closets and alcoves in the same way. Add the results.

If you buy foot-square tiles, simply convert the square footage of the floor into the number of tiles and add about 5 percent for cutting, waste, and possible future replacement (the last is very important).

If ordering tiles of any other size, multiply the total square footage of the room by 144; find the square-inch area of the tiles, then divide the first figure by the second. Add 5 percent to the answer.

Order adhesive along with the tiles. Since there are several kinds, use the one recommended by the flooring dealer for the type of tile as well as for the type and location of the subfloor. The amount required varies.

Preparing the subfloor.

One of the biggest questions in homeowners' minds is, "Can I lay a new resilient tile floor over an old resilient tile floor?" The manufacturers' answer is a qualified yes, mine is a qualified no.

Here, for example, is what the Kentile people have to say on the subject:

1) Kentile asphalt, vinyl asbestos and solid vinyl tile may be installed over suspended existing floors of asphalt, vinyl asbestos and solid vinyl tile and vinyl sheet flooring, providing the existing floor is smooth (not embossed) and does not have a foam or cushioned back.

2) Asphalt and vinyl asbestos tile may be installed over existing resilient floors of asphalt, vinyl asbestos and solid vinyl tile and vinyl sheet flooring, on or above grade, provided that the existing floor covering is securely bonded, is not embossed and does not have a foam back.

3) None of our tile products may be installed over existing linoleum or cork tile on grade. Asphalt tile is not to be installed over existing linoleum or cork at any grade level.

4) Do not install our tile products over any type of embossed or cushioned material.

All of which is slightly confusing. But what really concerns me about tiling over old resilient-tile flooring is that it's not always possible for the layman to identify the old flooring. It's also hard to tell whether the old flooring is really stuck down as well as it may seem to be. And finally, when you add a new layer of flooring on top of an old (no matter whether you're dealing with resilient tile, wood, or stone), you often cause a variety of small but annoying problems. In a kitchen, for example, you must either butt the new tiles to the sides of the cabinets, thus creating small, dirt-catching cracks; or you must remove the cabinets, slip the new tiles underneath, and thus change the level of the counters so they're no longer flush with, say, the range top.

Taking up old resilient flooring is messy but not as difficult as it looks. In most cases, all you need is a flat garden spade. Work this under the flooring and push hard. If this doesn't work well on asphalt or vinyl-asbestos tile, fan the flame from a propane torch across the floor to soften the adhesive.

When all the flooring is off, remove the adhesive from the subfloor with a disk sander—either a rented floor sander or the sander attachment for an electric drill. Brown linoleum paste is easily scrubbed off with water. The subfloor must then be prepared according to the pertinent directions that follow.

Laying resilient tiles directly on a finish wood floor is permissible only if the floor is laid on a subfloor and made of strips less than 3 inches wide. Remove wax from the floor. Scrape off paint (but not transparent finishes unless they are in poor condition) down to the wood. Sand the entire floor if necessary to remove the ridges at the joints. Fill holes with plastic wood. Then, if any bare wood has been exposed by sanding or paint removal, size it with shellac to keep the boards from cupping.

One other step that is occasionally called for in preparing a floor made of narrow wood strips is to fill large depressions caused by the settlement of the joists. This is done by troweling on a latex-cement underlayment like that used on concrete subfloors (see below).

All other wood floors, including wood subfloors, must be covered with a rigid underlayment of hardboard, plywood, or particleboard before tile is laid. Finish wood floors that are laid over subfloors and built of strips or boards more than 3 inches wide must be covered with ¼-inch underlayment. Parquet floors should be treated similarly.

If a floor is constructed with only a single thickness of tongue-and-groove boards—whether these be used as a finish floor or subfloor—it must also be covered with ¼-inch underlayment. If the boards are not tongue-and-groove, use ½-inch plywood underlayment.

In all cases, hardboard underlayment should be of material specially made for the purpose or of standard hardboard. Tempered hardboard should not be used. Particleboard underlayment can be used only if it bears a National Particleboard Association (NPA) stamp marked "CS 236-66 Type 1-B-1." Plywood underlayment can be made of any interior or exterior plywood that is smooth and free of voids on the top side. Most often used are softwood plywoods labeled "Underlayment INT-DFPA" or "C-C Plugged EXT-DFPA."

The underlayment panels should be laid perpendicular to the subfloor with a 1/64-inch space (the thickness of a matchbook cover) around all edges. Nail the panels with annular-ring nails spaced 4 inches apart in all directions through the center and along the edges. Drive the nails down flush or very slightly below the underlayment surface.

Wood floors or subfloors built on sleepers over grade-level or below-grade concrete slabs as well as those laid directly on such slabs should not be used as a base for resilient flooring unless you know beyond the shadow of a doubt that moisture never penetrates the slab from the ground.

Ceramic tile floors can be covered with resilient tiles if the mortar joints are flush with the surface. If they are concave, clean them thoroughly and fill with ceramic-tile grout.

Concrete is the most difficult type of subfloor to deal with because, although it is an excellent base for resilient tile, the moisture, alkilis, and special compounds used in its construction may attack the resilient tile or weaken the adhesive bond. New slabs in contact with the ground should be allowed to cure for two months before tile is laid, suspended slabs for one month. In addition, unless you are certain that the slab was constructed with conventional materials, you should sand it hard to remove curing compounds, hardeners, sealers, and so on, which may have been employed by the mason.

All slabs, new and old, should be treated for moisture content. To do this, spread a 6-inch-wide strip of linoleum paste on the concrete, and alongside spread a similar strip of the adhesive recommended for installing the resilient tiles. Over these lay a large piece of heavy polyethylene film or rubber matting and stick the edges down with package-wrapping tape. After seventy-two hours, if the linoleum paste is dry and the other adhesive adheres to the floor, the concrete is dry enough for the tile to be laid. If the adhesives haven't set, the installation must be delayed.

Other steps that must be taken in preparing a concrete slab are as follows:

1) Remove all paint on slabs in contact with the ground. On suspended slabs you need to remove only loose and scaling paint.
2) Clean the concrete thoroughly. Oil, grease, and wax are particularly important to remove.
3) To remove alkilis, scrub the concrete with water, then with a solution of 1 part muriatic acid in 9 parts water, then with water again. When the floor is dry, pour about a teaspoon of water on the floor in various spots. Into this sprinkle a couple of drops of a 3 percent solution of phenolphthalein dissolved in alcohol. If the spots turn red, pink, or purple, it indicates that the alkili has not been neutralized, and you should scrub the floor again with muriatic acid.
4) Fill holes and cracks in the slab. (See Chapter 17.)
5) Chip off or grind down high spots and surface roughness.
6) To bring depressions in the concrete up to level, trowel in latex cement, strike it off flush with the surrounding surface, and trowel the edges to feather thickness. Deep depressions should be filled gradually in ⅛-inch layers.

Other preparations.

Remove shoe moldings. If there aren't any, remove the baseboards.

Take out radiators, if possible, and floor grilles over registers.

Although the normal practice is to tile up to the edges of built-in cabinets and toilet bowls, these may be removed if you prefer to run the tiles in underneath (thus simplifying the fitting of tiles around them).

Finally, sweep the floor clean.

Installing lining felt.

Lining felt compensates for the movement of wood subfloors and thus helps to prevent seams from opening in the resilient flooring. It also makes resilient floors quieter and more resilient and simplifies their removal.

Although it is sometimes omitted, felt should be used over all underlayments and must be used on board floors. Lay the felt perpendicular to boards; it can be laid in any direction over underlayment sheets.

Cut the felt into strips fitting from wall to wall. Spread linoleum cement on the floor with a notched trowel, and roll the strips into this. Butt the edges of adjacent strips. Work out bubbles with a flooring roller or rolling pin.

If you are putting down a floor in two adjoining rooms with floorboards that do not run in the same direction, cover the joint between the floors with a 4-inch strip of canvas. Extend the canvas only 1½ inches to one side of the joint and tack it down. Leave the 2½-inch flap loose. Stick the lining felt down over the canvas.

If the two adjoining floors are not of exactly the same height, sand or plane down the higher floor before putting down the canvas.

Laying out resilient tiles.

The easiest way to arrange resilient tiles is to start with full-width tiles in a corner and lay them across the floor to the opposite walls. Thus you have to cut the tiles only along two walls. This method should be used, however, only if the cut tiles are more or less hidden from view by furniture, appliances, or plumbing fixtures.

For best effect, tiles should be laid from the center of the floor so the border tiles next to the walls are of approximately equal width.

Find the middle of the end walls in the main part of the room (in other words, ignore alcoves, bays, etc.). Stretch a chalked cord between these marks and strike a line on the floor. Then find the middle of the line and draw a line at right angles to this with a carpenter's square. Stretch the chalked cord over this new line from side wall to side wall and strike a line on the floor.

(If you don't have a carpenter's square, find the middle of the side walls and stretch the chalked cord between them. Then from the point where the cord crosses the chalk line, measure 3 feet toward either side wall and make a mark. Measure 4 feet along the chalk line toward either end wall and make a second mark. Then measure the distance between the marks. If it is exactly 5 feet, the cord and chalk line are at right angles to one another, and you should snap a line on the floor with the chalked cord. If the distance between the marks is more or less than 5 feet, adjust the chalked cord.)

Starting at the crossing of the two chalk lines, lay a row of tiles along the lengthwise line to the ends of the room. Then lay a row of tiles along the other line to the side walls. See the first drawing. You will now find that the tiles at points A and B are of equal width, and those at C and D are of equal width. If A and B are also approximately equal in width to C and D, you can go ahead and lay the entire floor as

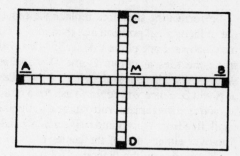

is. However, if there is considerable difference in the width of A and B on one hand and C and D on the other hand, take up the tiles, move the center tile half of its width to either side of the crosspoint, and re-lay the rows of tiles.

Suppose you're using 12- x 12-inch tiles, and the tiles at A and B are 10 inches wide while those at C and D are only 3 inches wide. Since this would not result in an ideal arrangement of the completed floor, move the center tile 6 inches toward either end wall, and re-lay all the tiles along the lengthwise line. Now you will find that the end tiles in the row—at points A and B—are only 2 inches wide—just 1 inch less than the C and D tiles. This is a much better arrangement than the first, since the border tiles around the entire room are almost the same width. Even so, the arrangement is far from ideal since the border is so narrow.

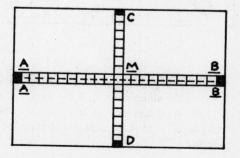

To correct the situation, replace the center tile in its original position and then move it 6 inches toward one of the side walls. This will make the tiles at points C and D 9 inches wide—only a little less than the 10-inch tiles at A and B. Since you couldn't ask for a more attractive arrangement, rub out the lengthwise chalk line from A to B and strike a new chalk line over either edge of the center tile.

To lay tiles on the diagonal, follow directions in Chapter 8 for laying parquet on the diagonal.

Lay two loose rows of tile from the center point to the walls as shown in the drawing. Lay one row flush to the diagonal lines. Lay the other row along line A-B with the points of the tiles touching. If the border tiles are not of appropriate width, move the tiles on the diagonal line half of their width to either side of the line.

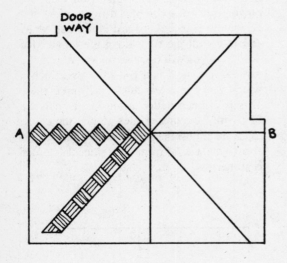

When laying diagonal tiles in two colors, try to use the same color half-tile against the wall all around the room. (If the half-tile is the wrong color, start at the center with the opposite color.)

Laying the floor.

Starting from the center of the room, tile one-quarter of the room at a time. Lay two or three tiles out along the intersecting chalk lines and fill in between them. Then build out from these in a gradually widening fan.

The great majority of tiles are laid with the pattern at right angles to the pattern in adjacent tiles. Usually the pattern is obvious. If not, arrows are printed on the back.

How much floor should be covered with adhesive depends on the type of adhesive used. If the adhesive dries rapidly, spread it out over 6 to 10 square feet on the subfloor and lay the tiles in it immediately. If the adhesive is supposed to become tacky before the tiles are laid, you can spread it over at least one-quarter of the floor. Follow the directions on the can.

To set a tile, slant one edge to the floor tight against the chalk line or preceding tile, drop the other edge, and press the tile firmly into the adhesive. Never slide a tile.

If the adhesive is of the slow-setting type, you can walk or kneel on the newly laid tiles at once. They should not slip out of place. With fast-setting adhesive, however, stay off the tiles until they no longer move when you push them with a finger. If this is impossible, cover them with plywood and work from this.

Solid vinyl, rubber, and cork tiles should be rolled after you have laid a dozen or so. Use a heavy floor roller borrowed or rented from the flooring dealer or an ordinary rolling pin.

Vinyl-asbestos and asphalt tiles are too brittle to be rolled. After you have covered a sizable area, just go back over them and press them down with your hands. It pays to do this several times.

Adhesive that oozes out between joints should be wiped off immediately. Then, when the entire floor is laid, go over it once more with water or the solvent specified by the adhesive maker.

Although some tiles are thin enough to be cut with scissors, you can make a cleaner cut with a linoleum knife that has a hook-shaped end and a sharp point. Hold the tile face up on an old piece of plywood. For straight cuts, lay a steel straightedge across the tile and run the knife along the edge. Don't cut too fast: The knife may slip. For rounded cuts, draw a pencil line on the tile and cut—slowly—along it.

To fit tiles to a wall, follow the directions for fitting parquet flooring (see Chapter 8). Fitting around pipes, radiator legs, toilets, bathtubs, and other irregular objects is done by drawing a pattern on paper. Cut this out and try it in the space to be filled; then, if it fits neatly, transfer the pattern to the tile. Small gaps around the cut edges of the tile when it's installed can be filled with bathtub caulking or colored plastic wood.

If a tile fits completely around an object such as a pipe, cut out the required hole and make a straight cut from it to the nearest edge. Bend the cut edges back just far enough to go around the object. Take pains not to tear or break the tile.

If tiles are laid under a radiator instead of around the legs, place steel nuts ⅛ inch thick under the legs so the radiator won't rest directly on the flooring. Hammer the nuts into the tiles (after heating asphalt and vinyl-asbestos tiles).

Problems with asphalt tiles.

Because asphalt tiles are very hard and brittle, they are difficult to cut and mold to slightly uneven subfloors. Straight cuts are best made by drawing your knife across them repeatedly and then snapping in two. For irregular cuts, place the tile in a low oven until it is pliable. Don't let it heat too much. Then cut it immediately. (Vinyl-asbestos tiles can also be heated for easier cutting.)

You should also heat asphalt tiles slightly to make them conform to an uneven subfloor. The alternative is to heat them with a propane torch after they have been glued down. Just be careful not to get the flame too close or hold it on a tile too long.

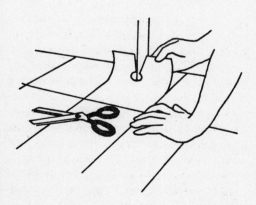

Laying a floor with feature strips.

Feature strips are long, narrow strips of resilient flooring that are inserted between rows of tiles (or sheets) to give extra interest to the floor. When using them, be sure to work out the overall design of the floor before starting to lay tiles. Do this on paper and then lay the tiles and strips out loosely on the subfloor.

Strips are laid like square tiles. Despite the temptation to do so, never slide them into place. Set them exactly where you want them and smooth them down. When laid, the strips should be adjoined by tiles for their full length; if they extend out into the adhesive by themselves, they may be damaged or stuck down in a bent position.

Installing accessories.

There are several accessories that you may need when covering floors with resilient tiles:

Edgers are metal or vinyl strips that are used to protect the edges of tiles that project above an adjoining floor (for example, when you lay tile over an existing floor which is level with the dining room floor). Some edgers are more or less L-shaped metal strips that are nailed to the floor. The tiles are then laid over the nailing flange behind a raised protective nosing. Other edgers are simple vinyl strips that are glued to the floor adjacent to the tiles.

Step nosings made of vinyl or metal perform the same function as edgers when resilient tiles are laid along the tops of steps.

Wall bases are flexible vinyl or rubber ribbons widely used in kitchens and bathrooms in place of wood baseboards. So-called straight, or carpet bases are flat strips that are cemented to walls just above resilient floors; cove bases—more common—are rounded outward at the bottom to eliminate sharp, dust-catching corners and seams between floors and walls. The bases come in many colors and finishes and in heights ranging from 1½ to 7 inches.

Walls on which a base is used must be sound—without breaks or gaps—all the way to the floor. Clean well.

The bases can be run all the way around the room in a continuous strip or cut into lengths and installed between preformed corners. The latter method is much easier. In either case, installation is made by applying wall-base adhesive to the back of a strip with a notched trowel and pressing the strip against the wall. Roll it down firmly with a Mason jar or wallpaper roller, and press the curved toe against the floor with a straight piece of wood.

Laying self-adhering tiles.

Self-stick tiles are made strictly for people who want to improve the appearance of a floor with the least possible work at lowest possible cost, and as long as the floor lasts until they move to another house, that's plenty good enough.

In other words, if you want a floor to give good service over many years, don't buy tiles of this kind. They're made of thin vinyl-asbestos that is inclined to tear and crack when you handle it. The adhesive that comes applied to the backs of the tiles has about as much stickum as a twice-used postage stamp. And the theory that the self-adhering feature saves you lots and lots of work is nothing but advertising blarney.

The truth is that the only thing different about putting down self-stick tiles is that you don't have to spread adhesive on the floor. You simply peel off the protective paper on the back of the tile and press the tile directly onto the subfloor.

Finishing asphalt, vinyl-asbestos, and vinyl-tile floors.

You might think that by now every American housewife knows that she should never, never apply a hard finish to resilient flooring. But there are still thousands who don't, and sooner or later they write to someone like me complaining, ''What should I do? I put lacquer on my kitchen linoleum to keep it looking shiny. The first time it began to wear thin, I just put on another coat. But I'm not sure what I should do now. It's wearing off again in the traffic lanes—worse than before. And I notice that it's turned sort of brownish-colored in the other areas.''

Well, lady, you've worked yourself into a position from which there is no return. The only way to remove the hard finish is with a solvent that will wreak havoc with the flooring. And every time you touch up worn spots with new finish, the overlapping areas get a little darker.

The only finish that should be used on resilient floors is water-based wax or a clear acrylic finish. (On cork flooring, however, use solvent-based wax or acrylic.)

Three types of wax are available.

Self-polishing waxes are most popular because you just wipe them on a floor with a wool applicator and forget them. Drying to a tough, scuff-resistant film with a bright shine, they require no buffing (and, in fact, they don't look well if they are buffed). When they finally become worn, the entire floor must be rewaxed. But because repeated waxing just increases the thickness of the film, floors finished with self-polishing wax must be stripped fairly frequently.

Buffable waxes must be buffed with a floor polisher to achieve a bright luster, and they show scuff marks. But when they begin to look sad, instead of adding a new layer of wax, you just give them a good buffing and they're as good as new.

Clean-and-polish waxes are self-polishing waxes containing a detergent that cleans as the wax is being applied. Thus, they save time by eliminating separate washing-rinsing operations. But unfortunately, they are not very durable.

Acrylic floor finishes are self-polishing materials that are applied with a wool applicator. The resulting finish looks like a self-polishing wax finish but lasts longer.

Whichever kind of finish you prefer, wait at least three days after the floor is laid before making the first application. Thereafter, apply wax only when the floor really needs it, and when you do wax, don't apply it under the overhangs of base cabinets or within 6 inches of walls, because these areas get little wear.

Maintaining asphalt, rubber, vinyl-asbestos, and vinyl floors.

Follow this schedule:
1) Sweep or vacuum the floor daily to remove grit that will make scratches. Wipe up spills at once. Remove heel marks with a heavy-duty detergent or white appliance wax.
2) Mop once or twice a week with a well-wrung-out mop.
3) Mop every three or four weeks with a dilute detergent solution and rinse. Use a damp—not wet—mop both times. Too much washing is not good for resilient floors.
4) Apply a very thin coat of wax after every third or fourth washing with detergent.
5) Strip the entire floor when the buildup of wax begins to yellow and look uneven. If you have cared for the floor properly, this operation shouldn't be necessary more than once a year. If you're uncertain whether stripping is called for, make a test on an inconspicuous area. A marked improvement in the color of the floor indicates you should do the entire floor.

Stripping is most easily done with a one-step packaged wax remover, although you can substitute a solution of ammonia and water. Sweep the floor first. Dissolve the wax remover in water and slosh it on a 10-square-foot area. Let it stand for about three minutes. Then scrub with an electric floor scrubber, stiff fiber brush, or fine steel wool. When all the wax is off, sponge up the mess and do another area. Rinse thoroughly when the entire floor is clean, and look for areas you may have missed. Strip these once more and rinse. After the floor dries, wax it.

Removing stains from asphalt, rubber, vinyl-asbestos, and vinyl floors.

Try a mild detergent solution first. If this doesn't work, use a white appliance wax. It is particularly good on food stains, heel marks, grease, oil, tar, shoe polish, and crayon. If the stains persist, scrub with very fine steel wool and household cleanser.

For burn marks use cleanser and steel wool.

Solvents should be wiped off at once. If they harden before you get to them, scrape off with a dull knife. Then rub with fine steel wool and household cleanser.

Scrape off paint and chewing gum with a dull knife. Then rub with trichloroethylene.

Rust is removed with a cleanser containing oxalic acid—Zud, for example.

Some ink marks will come off with water, so try this first. If the stains remain, cover them for a few minutes with a cloth dipped in alcohol, then scrub. Repeat the process with household ammonia as necessary.

Mustard stains are extremely hard to remove, so spills should be wiped up at once with a pick-up rather than spreading motion. Then wash with detergent solution. If discoloration remains, wash with 3 to 5 percent hydrogen peroxide and let stand for a few minutes before rinsing.

Photographic chemicals are even more difficult than mustard. If you use them regularly in a room with a resilient floor, write the flooring manufacturer for information on specific reagents.

Repairing asphalt, rubber, vinyl-asbestos, and vinyl flooring.

It's best to leave scratches alone: They tend to disappear in time. In the meanwhile, you can pretty well obliterate them by applying wax. If not satisfied with this treatment, remove scratches in asphalt and vinyl-asbestos floors by sanding with fine sandpaper and waxing. On vinyl and rubber floors, rub scratches lengthwise—not too hard—with a worn screwdriver blade or an old coin. Because the material is fairly soft, this compresses the edges of the scratches, leaving only a thin line. Use sandpaper only as a last resort.

Small holes in vinyl can be repaired by scraping a scrap of vinyl flooring into dusty particles and mixing with a little acetone or methyl ethyl ketone to form a putty. Add a few drops of lacquer or clear fingernail polish. Stick cellophane tape down around the hole, and fill the hole with putty to the level of the tape. Smooth it with a putty knife. After the putty has set for a few minutes, peel off the tape. As the putty dries, it will shrink down to the level of the floor.

There is no way to repair holes in asphalt, rubber, and vinyl-asbestos tiles.

Dents in resilient flooring other than rubber can't be eradicated. Sometimes they repair themselves in solid vinyl, but don't count on this. The only way to prevent dents is to use large, flat-bottomed floor protectors or furniture cups under legs of furniture. Several types are illustrated.

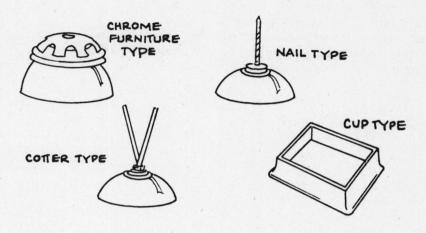

CHROME FURNITURE TYPE

NAIL TYPE

COTTER TYPE

CUP TYPE

If a tile is cracked, broken, or has a large hole, it must be replaced. But before you rip out the defective tile, make sure you can get a new one of the same color and pattern. Unfortunately, the odds are against this, because resilient flooring manufacturers no longer keep old patterns very long. This is why, when you buy tiles for a new floor, you should order two or three extras to stow away for future repairs. When you dig them out of storage, they may not be exactly the same color as the floor, which will have faded, but they will be better than nothing.

To remove a damaged tile, pry it out with a chisel or stiff putty knife. Always work from the break toward the edges of the tile, because if you work from the edges in, you will probably damage the surrounding tiles. When the tile is up, scrape as much of the adhesive as possible from the subfloor. Then test whether the new tile will fit into the hole. It may be necessary to scrape down the edges very slightly with a sharp knife. Then apply adhesive to the back of the tile and press it down.

Finishing, maintaining, and repairing cork tiles.

Normally prefinished and unfinished cork are given only a couple of coats of solvent-based paste wax. If an unfinished cork floor is subject to considerable staining, however, it's advisable to apply two coats of a clear penetrating floor sealer like that used on wood floors. Apply paste or buffable liquid wax over this.

Treat a cork floor like wood. Avoid washing with water. Just vacuum frequently and buff periodically. When it begins to look dirty, an application of a thin coat of wax will spruce it up. When very dirty, strip off the wax with a waterless floor cleaner recommended for wood, then rewax.

Fill holes and bad dents by grating a piece of bottle cork on a kitchen grater and mixing with white shellac to form a paste. Immediately spread this into the holes, and sand smooth when dry.

Badly damaged tiles must be replaced with

new ones. The latter will undoubtedly be much darker than the rest of the floor, but will fade rather rapidly upon exposure to light.

For stains there is no sure cure. Wash with detergent solution. Rub with white appliance wax. If these measures fail, your only hope is to remove the stains by careful sanding with medium and then fine sandpaper, but you may sand through the tile before the last discoloration is gone.

You can also remove burns by sanding, but unless the char is shallow, you'll probably do better to treat it like a hole or dent.

11/ RESILIENT SHEETS

The best resilient flooring is produced in sheets that come from the factory in enormous rolls.

Sheet patterns are more beautiful than tile patterns. The quality of the best sheets is superior to that of the best tiles. Some types of sheet have a layer of cushioning which makes them exceptionally quiet and comfortable underfoot. And because the sheets are laid with few if any joints, the resulting floor is virtually leakproof and especially easy to clean.

Naturally, the cost of sheet flooring reflects these advantages. You can't buy it for much less than 75 cents a square foot. And in the old days, unless you were adventuresome, you could expect the total cost of a sheet floor to be about twice the cost of the material alone, because you had to have it laid by a flooring contractor.

Most people today still depend on professional installers. But this dependence is weakening. Even the flooring manufacturers have seen the handwriting on the wall and have finally—and I think reluctantly—started to publish instructions for how to lay sheet floors yourself. Note, however, that these instructions are limited to 12-foot-wide sheets (which fit all the way across many rooms) of rotovinyl or new ultraflexible cushioned vinyls like Armstrong's Tredway. Don't try to lay any other type of sheet material yourself. Even with schooling, a good many professionals do the job badly.

Types of sheet flooring.

Linoleum. I'll start with this old favorite because even though it is slowly fading away it is still an outstanding material. Its wearability is legendary; its resistance to grease and burns and its ease of maintenance are outstanding. But it is made in only a limited choice of colors and patterns—most of them rather drab. And it can be installed only on suspended floors.

Solid vinyl. Vinyl flooring was linoleum's nemesis. Ever since its introduction, it has captured a bigger and bigger share of the market. This is because no other material has so many plusses and so few minuses. Solid vinyl sheet has beauty and durability. It's exceptionally easy to keep clean and maintain. It resists staining and the harmful effects of acid, alkilis, grease, and most household chemicals. And it's just as pleasant to walk on as linoleum, which has always been one of the quietest and most resilient floorings.

In fact, the only real black mark against solid vinyl—as well as all other kinds of vinyl—is its poor resistance to burns.

An additional advantage of solid vinyl flooring is that it can be professionally installed over old resilient flooring as well as other old and new subfloors by Armstrong's Perimiflor installation system. This differs from the conventional installation system in that the adhesive is spread in a ribbon only around the perimeter of the floor and at seam lines and fixtures in between. While this doesn't actually cut the installation time, it is supposed to give the installer more time to fit the flooring to produce a tighter, better-looking finish floor.

Cushioned vinyl. This is a solid vinyl sheet bonded to a thick layer of vinyl foam, which gives as you walk across the floor. The result is more carpetlike than any other built-in flooring material—including cork. Not only is the material extremely comfortable underfoot, but it is also very quiet. If you have a downstairs room that echoes with the noise of people trotting around upstairs, this is an ideal flooring to install in the upstairs rooms.

Although cushioned vinyl is one of the thickest and also one of the most expensive resilient flooring materials (prices start at about $1.60 a square foot), it is not as durable as many less expensive solid vinyls. The actual wearing surface is quite thin.

Rotovinyl. Rotovinyl is a fairly inexpensive sheet consisting of a felt or asbestos backing to which a design is printed by the rotogravure process. A protective coating of clear vinyl is applied on top. As a rule, the sheet also contains a core of vinyl foam to provide greater comfort.

Some rotovinyls can be laid only on suspended floors; others can be laid on any grade level. Even the latter, however, are poor substitutes for a real vinyl floor because they have a very thin wearing surface.

Shiny vinyl. Vinyls with an ultrahigh-gloss finish that requires no waxing are the latest flooring rage, and that goes to show how far out in left field I am about some things, because I think they are hideous. The colors are lovely, the patterns attractive. But that mirrorlike shine is just too much. It's completely artificial, doesn't have any place in a house, doesn't do anything for the house. When the sun's shining on it, it is actually uncomfortable to look at.

Another objection to the flooring is that many people buy it in the belief that it will be so much easier to care for than ordinary vinyl. But the truth is that, except for not needing waxing (in fact, wax won't stick to it), it must be swept and mopped just as often, and even with constant care, the shine doesn't last forever in traffic areas. It may, indeed, disappear within a few months if you have a large family. Then you must go downtown and buy a can of special floor finish to restore the luster.

Do-it-yourself vinyl. This is a new kind of flooring with a vinyl wear-layer, foam inner-cushion, and special backing. It's made only in 12-foot widths and is so flexible that it can be folded like a blanket. It is also so elastic that it stays taut even if the subfloor expands and contracts. Installation is made with staples or adhesive around the perimeter of the sheet. However, the material lacks the durability of other sheets and is presently available in a very small selection of colors and patterns.

Rubber. Rubber sheet is similar to rubber tile and even less common. Available in 3/32-, ⅛- and 3/16-inch thicknesses, you might find it especially suited to your family room not only because of its resilience but also because it resists cigarette scorching. The wearing surface extends from top to bottom. The rolls are only 3 feet wide, however, which means more seams in a floor.

Estimating how much sheet material needed for do-it-yourself installation.

Sheet flooring is like wallpaper, in that some designs have a definite pattern that must be matched from sheet to sheet, and all pieces should be laid in the same direction. This makes it difficult to estimate how much material you need, because in a room requiring two strips, one piece may have to be longer than the other to permit pattern-matching. Furthermore, you can't chop off the end of a sheet and turn it sideways to fill in, say, an alcove next to the whole sheet. For these reasons, the only way to order the flooring you need is to draw an accurate plan of the room, take it to the dealer, and ask him to supply the answer. He will also advise how much adhesive of which type you should have.

Preparing the room.

The subfloor should be prepared like one to receive resilient tiles (see Chapter 10). A slight difference is that, if you are loose-laying a sheet flooring, it's unnecessary to remove old resilient material unless it is a cushioned vinyl. But you must fill all holes and depressions in the old floor with latex cement or by cutting out around them and cementing in scraps of flooring. You must also sand the flooring wherever the new material is to be cemented down.

Remove shoe moldings or baseboards so the new flooring can be laid to within ¼ inch of the walls. Take out radiators. Sweep the floor clean.

Laying a sheet floor in adhesive.

Unroll the flooring in a large room where it can lie flat. The room—like that in which the floor is to go down—should be warm enough to allow the sheet to relax and gain flexibility.

Draw a precise plan of the room in which you're laying the floor. Since the flooring should as a rule run lengthwise in the room, measure 3 or 4 feet out from one of the long walls. Make the measurements at the end walls, stretch a chalked cord between the marks, and snap a chalk line on the floor. Then strike a second chalk line all the way across the room. Use a carpenter's square so the two lines are at exact right angles to each other.

Working from these two lines, measure the walls, base cabinets, built-ins, alcoves, etc., all the way around the room. Make the measurements very carefully and double-check them. Be sure to measure not only the length of the walls, and so forth, but also the distance that base cabinets, etc., project out from them. As you make the measurements, transfer the information to a paper plan of the room.

When the plan is finished, examine the new sheet flooring and decide which of the two long edges will be placed against the lengthwise wall from which you started making measurements. If the edge was damaged during delivery, lay a straight board parallel to it and cut it off with a sharp linoleum knife or utility knife. (Always place a board or piece of plywood under the sheet when cutting it; otherwise you will damage the floor beneath or dull the knife unnecessarily.)

Strike chalk lines on the sheet corresponding to those in the room to be floored. Then, working from your plan, draw an exact outline of the floor on the sheet with a soft pencil. Recheck your measurements once more. Cut the flooring along the lines.

Roll the sheet crosswise into a loose but manageable bundle that you can roll out lengthwise in the room to be floored. The back of the sheet should face out. With a helper, carry the sheet to the room, roll it out on the floor loosely—without forcing it into offsets and under base cabinets—and check the fit. Then fold half of the sheet back on itself, and cover the subfloor exposed with adhesive. Embed the sheet in this and glue down the other half. Finally, roll the floor with a rented floor roller or rolling pin. Be certain to eliminate all air pockets.

This process sounds simpler than it is, because in kitchens and bathrooms you almost always have problems slipping the flooring into position under overhangs, around corners of cabinets, and into other tight places. Unfortunately, I have no magic directions for simplifying the job. In some cases, you should roll the flooring back slightly and ease it into position. In other cases, you may have to put a belly in the sheet (as an inchworm humps himself across a branch), slide the end under the obstruction, and smooth the belly down. Whatever procedure you follow, don't tear or crack the sheet.

To work a sheet around a pipe, lap it vertically against the pipe and push it against the base of the pipe as closely as possible. Then make a straight cut down through the lap over the middle of the pipe to the floor. Make a short horizontal cut around the bottom of the pipe. Fold the sheet around the pipe, press it in at the bottom, and cut it out bit by bit around the pipe until both flaps lie flat on the subfloor.

Fitting sheet material tightly around door casings is another difficult task, especially if the casings are contoured. One procedure is to lap the sheet up against the casings and cut through it on either side of the casings. Then pare off the lapped sections bit by bit until the sheet can be pressed to the floor. Another somewhat simpler procedure is to cut a fraction of an inch off the bottoms of the casings before spreading adhesive on the subfloor. To do this, lay a scrap of flooring alongside each casing, place a fine-toothed saw flat on this, and saw through the casings. Rake out the chips and sawdust. Then when you glue the sheet, make a slight belly in it and push the edge under the casings.

When the flooring is down and rolled, replace the shoe moldings or baseboards. First, place these loosely in position and check whether the flooring edges are covered. If they're not, cut a narrow strip of flooring and fit it between the large sheet and the wall. Then nail the moldings down. The not-so-good alternative is to replace the shoe moldings with the next larger size. If the room didn't have shoe moldings, an easy way to handle gaps between the flooring and baseboards is to install moldings.

In doorways, protect the edges of the sheet by installing metal or vinyl edger strips. Use step nosings to protect the sheet at the top of steps. (See Chapter 10.)

Allow the adhesive to dry for twenty-four hours before permitting traffic on the floor.

If you are covering a floor more than 12 feet wide at any point, two strips of sheet material must be installed. To achieve a perfect fit between these, lay them together on the cutting floor before transferring measurements to them. Match the pattern (if required). Then overlap the edges of the sheets. Center a steel straightedge on the lap and use it as a guide to cut through both sheets at the same time. Remove the scraps and butt the sheets together. They will fit perfectly.

Place a strip of masking tape over the joint to hold the pieces together. Then draw the outline of the floor on the sheets and cut them on these lines.

Remove the masking tape and transfer the sheets to the room to be floored. Roll out the larger sheet first and position it. Then roll out the smaller sheet and position it alongside. Replace the masking tape at one end of the floor. Then fold back the sheets at the other end halfway and spread adhesive on the sub-floor. When both sheets are firmly embedded in this, remove the masking tape once more; fold back the second half of the flooring and stick it down.

Laying a do-it-yourself sheet floor.

Since this material is so flexible, the installation procedure is considerably simplified. There is no need to measure the room and cut the flooring in another room.

After removing the shoe moldings or baseboards and cleaning the subfloor, roll out the sheet on the floor, work out the bulges, and slide it into position. Place one edge tight against one of the long walls; let the other edges lap up the other walls. Make vertical cuts at the corners so the sheet lies flat on the floor at these points.

Starting along one wall, crease the sheet into the corner between floor and wall with a carpenter's square, and cut off the excess material lapping up the wall. Remove enough so the cut edge lies flat on the floor. If there's a slight gap between it and the baseboard, no harm is done. Around casings, however, the flooring should butt tightly—without curving up the casings.

When the entire sheet has been trimmed, fold it back from doorways and tops of steps and apply a narrow band of adhesive to the subfloor with the applicator supplied by the flooring manufacturer. Press the edge into this immediately. Then fasten the other edges of the sheet with a staple gun loaded with ⅜- or 9/16-inch staples. Hold the gun flat on the floor and drive in the staples about ¼ inch from the edges. Space the staples 3 inches apart. Take care throughout this cementing and stapling operation that you don't shift the sheet or allow it to develop bulges.

Finally, replace the shoe moldings or baseboards and protect exposed edges of the sheet with metal or vinyl edgers. The floor is now ready for service.

If it's necessary to use two strips of do-it-yourself sheet material to cover a floor, stick a strip of double-faced carpet tape to the subfloor under the joint and roll the strips into this. The joint is then sealed with cement supplied by the manufacturer.

Finishing and maintaining resilient sheet floors.

If conventional flooring is used, follow the directions for finishing and maintaining resilient tile (see Chapter 10).

Shiny vinyl floors should be swept or vacuumed even more often than other sheets, since scratches made by grit show up with exceptional clarity. Damp-mop with water or a mild detergent solution (never soap). Put enough pressure on the mop to get into surface indentations. For stubborn soil, scrub with a nylon cleaning pad (never with steel wool).

When shiny vinyl becomes extremely dirty overall, strip it with a one-step packaged wax remover applied full strength, and rinse well. Placing mats or rugs at outside entrances helps to reduce all these chores. But never use rubber-backed mats, since they may stain the flooring.

When shiny vinyl is badly scratched or begins to lose its gloss, wash and dry it; then apply the special vinyl dressing made for such floors by the flooring manufacturer. (As a rule, one manufacturer's dressing can be substituted for another's. But just in case you should have real trouble with the floor, it's wise always to use the dressing sold by the maker of the flooring; then there will be less argument about whether you maintained the floor properly.) Don't just slosh the dressing on the floor and spread it around. Pour it into a shallow pan into which you can dip the wool applicator used for applying the dressing.

Apply in thin, even coats.

The frequency with which you must dress a floor depends on the traffic over it.

Removing stains.

Follow the directions for stain removal in Chapter 10. But don't use alcohol on shiny vinyl, because it softens the surface. (By the same token, wipe up alcohol spills at once.)

Repairs.

Fix scratches and small holes as in resilient tiles. Handle other problems as follows:

To patch a large hole, place a scrap of flooring over the hole; hold it down firmly and cut through both layers at once. Then scrape out the defective flooring and old adhesive, and glue the patch with the appropriate adhesive.

If a floor develops a bulge, make a clean cut through the center of the bulge and spread adhesive underneath with a table knife or artist's spatula. It may then be necessary to shave down one of the cut edges in order to make the flooring lie flat. Weight down under a stack of books.

Tears in linoleum are treated in the same way. If you can lift the torn edges without aggravating the tear, scrape out as much of the old adhesive as you can before spreading in the new.

Torn vinyl is also reglued. Then to seal the edges together and conceal the tear to some extent, lay a 1-inch strip of smooth, heavy-duty aluminum foil over it and fasten one end with masking tape. Smooth the foil with your finger. Heat a hand iron to highest temperature and pull the point over the foil several times. Don't bear down or hold the iron in one spot. Wipe the foil with a damp rag and pull it off. Then clean the vinyl with water and household cleanser to remove the dull streak left by the foil.

12/ SEAMLESS FLOORING

Strange things have happened to the seamless flooring business. Just ten years ago, to hear industry members talk, you would have thought seamless floors, which are poured from cans and applied with rollers and/or trowels, were about to take over the world. You wouldn't have given a nickel for the future of resilient flooring—seamless's main competition—or, for that matter, anything except wood.

Then suddenly—or maybe it only seems suddenly in retrospect—silence. The bubble burst. I've heard various explanations. The most plausible is that the installed floors were fairly high in cost—which scared off consumers—and they took a couple of days to install—which ran up labor costs for the installers. It's also said that a lot of installations failed because they were poorly made. And there is no doubt that the lack of patterns and colors discouraged interest by homeowners, who have been on a pattern and color binge.

Despite the comedown, however, seamless flooring is still around. Perhaps it is even beginning to make a little comeback. But it's a leadpipe cinch that, having missed its big

chance in the sixties, it no longer offers a very serious threat to the resilient manufacturers. Why? Well, without going into the advantages of resilient, consider the drawbacks of seamless:

There is that limited choice of patterns and colors I already mentioned. This doesn't happen to bother me: I like the kind of nonpatterned, nubby, or flaky texture—something akin to a pepper-and-salt carpet—you get with seamless flooring. It has enough character to be interesting but it doesn't dominate any room in which it's used. It looks to me as a floor in a kitchen or bathroom or possibly a family room used by kids should look: attractive but serviceable.

Installation is another problem. At best it takes about three days and involves so many steps that the opportunities for error are multiplied. Furthermore, because the floors are seamless—poured in a single room-size sheet—you can't make any mistake in preparing the subfloor because you're almost certain to wind up with breaks or weak spots in the finish floor. All of which explains why professional installers with indifferent labor crews had and have trouble producing top-notch floors. But this need not mean that the do-it-yourselfer must experience the same difficulties. In fact, if you follow the manufacturer's directions and proceed with care, you should not.

Despite these negatives, seamless flooring has several definite plusses:

First, as its generic name indicates, it is completely lacking in joints or seams through which water, or anything else, can seep. This also makes it easier to clean.

Second, it shouldn't be waxed—which means that you can maintain its luster with little work. I have heard of one family that has had seamless flooring in the front hall for about eight years, and they still haven't had to reglaze it.

Finally, as the preceding sentence indicates, the flooring is quite resistant to wear as well as to chemicals and indentation. But despite claims, it is not very resilient.

Cost of the materials comes to about 55 cents a square foot.

Estimating your needs.

The amount of flooring required depends on the make and type, so the only way to figure how much you should buy is to measure the square footage of the floor and ask the dealer for the answer. In all cases, you require a primer, or base, coat, an aggregate consisting of vinyl chips or flakes, a topcoat, and a finish coat. You may also need an underlayment felt.

Preparing the subfloor.

Seamless flooring can be laid over concrete, plywood, particleboard, or ceramic tile. The subfloor must always be sound, clean, and level (you can't build up the thickness of seamless flooring enough to compensate for any appreciable depressions in the subfloor).

Concrete slabs may be suspended, on grade or below grade, but the latter two must be laid over a polyethylene moisture barrier. Remove all paint. Then to get off dirt, stains, alkilis, and so forth, you should either go over the floor with a terrazzo grinder or etch it with 1 part muriatic acid in 4 to 5 parts water. First dampen the floor with water, then scrub it thoroughly with acid and rinse well. Then flood the floor with a strong solution of household ammonia, let stand for about ten minutes, rinse, and dry. Fill holes, cracks, and low spots with latex cement. If a great deal of patching is required, apply a rubber-asbestos underlayment felt such as Lexide SA-305.

Ceramic tile should be thoroughly cleaned with a household cleanser and rinsed. Remove loose tiles and mastic and fill the voids with latex cement.

Plywood and particleboard are handled in two ways after you've made sure the sheets are firmly nailed to the joists. The joints are either sealed by covering them with fiberglass tape embedded in epoxy adhesive or the entire surface is covered with rubber-asbestos underlayment felt. The second method, which is the newer, is in many ways simpler and yields a better surface for the seamless flooring. Put down the underlayment with the adhesive recommended by the manufacturer. Spread this over the subfloor with a notched trowel and roll the felt out into it. Butt the edges of adjacent strips carefully. Go over the entire floor with a resilient flooring roller or a rolling pin to eliminate every suspicion of a bubble. The seamless flooring can be installed after an hour's wait.

Installing seamless flooring.

The installation procedure varies, so follow the manufacturer's directions. Some flooring is put down with a trowel in a layer about ⅛ inch thick. I strongly recommend that you use the type that is rolled on.

Since seamless flooring is made of epoxies and other irritating chemicals, the room must be well ventilated during installation (one maker even recommends the use of a safety mask with respirator for one step of the process). The room temperature should be no lower than 50°—preferably 70°.

Wear white-soled shoes or plastic booties.

Apply the primer with a short-napped paint roller and take pains not to skip areas, especially near the edges of the floor and around built-in fixtures, pipes, and so on. Work out ridges in the primer and don't leave puddles. Apply only as much primer as you can spread and cover with granules within about thirty minutes.

Immediately sprinkle the primer with granules or chips as supplied by the manufacturer. Don't mix sizes. Distribute the aggregate as evenly as possible, but don't worry if you get more in some places than in others. The important thing is to cover the floor until the primer is no longer visible.

Let the floor dry for six to twenty-four hours, according to directions, until it is hard enough to walk on. Then sweep off the excess granules or chips, and vacuum up the dust. Finally, scrape the floor with the edge of a trowel to remove projecting pieces, and vacuum once more.

Now apply the topcoat (or coats) with a rubber squeegee followed by a short-napped roller. The floor should be well wetted with the topcoat but not flooded. Roll in all directions to ensure even application.

After the topcoat has dried overnight, sand the entire floor to remove sharp granules and smooth out bumps. Sanding may leave some scratches, but these will disappear during the final application step. Vacuum the floor clean and inspect it for thin spots. If you find any of these, cover them with topcoating, dry, sand, and vacuum again.

The finish coat is a combination sealer and glaze which gives the floor a uniform luster. The number of coats required depends on the make of flooring and also on the expected traffic. Apply with a short-napped roller in thin, even films. Roll in all directions.

Let the floor cure for at least twelve hours before walking on it—and don't walk on it too much. Twenty-four- to thirty-six-hour curing is required before the floor is ready for normal traffic.

Maintaining seamless flooring.

A finish seamless floor does not require waxing. In fact, waxing is usually inadvisable since it tends to pick up dirt, and this interferes with floor cleaning. Just sweep the floor daily and mop with water or mild detergent solution when it is soiled. If the floor has a semirough texture, use an electric floor scrubber.

Eventually, when the floor shows wear, it should be reglazed. Scour the floor with household cleanser and rinse thoroughly. When dry, go over it with very fine sandpaper or medium steel wool and vacuum up the residue. Then roll on one or two new coats of the sealer and glaze coat. Let these dry for twenty-four hours or more before using the floor.

Repairs.

Scratches in seamless floors are "removed" by reglazing. Sand them lightly and then simply brush or roll on the same material used to give the floor its final finish.

Holes and other damaged areas can be repaired only by redoing them more or less from scratch. If the damage is not deep, you may get by simply by sanding down to the base coat of chips; in other cases, you may have to go all the way to the subfloor. Then fill in the hole as if you were installing a new floor. Unfortunately, since the patch probably won't blend into the surrounding area, you must finish the job by reglazing the entire floor, scattering granules over it, sanding, and applying one or two additional coats of glaze.

13/ CERAMIC TILE

I've just been leafing through the ceramic tile catalogs I've accumulated in the past couple of years. It was like looking at the Christmas catalogs from Neiman-Marcus, Gump's, and Tiffany. There are so many gorgeous tiles on the market today that you really can't decide which you want.

I can remember when the motto of ceramic tile manufacturers might well have been an adaptation of Henry Ford's famous saying about the Model T: "You can have any color ceramic tile you want as long as its white or black."

How things have changed! The tile people now offer so many colors you might think they invented color. Yes, and they have lots of different finishes and lots of different textures and reliefs, and lots of different shapes, and lots of different sizes. If you plan to lay a floor of ceramic tile, you had better set aside about a week to decide what you'll use.

Confusing as it is, this plethora of tiles has one thing in its favor: It gives you ideas. You begin to see that ceramic tile need not—and should not—be restricted to bathroom floors. It's also great for kitchens, laundries, halls, family rooms, dining rooms, lanais— wherever you want bright color, interesting pattern and texture, and the practicality of a waterproof, stain-resistant, easy-to-clean floor—and never mind the fact that it also happens to be cold, hard, and semislippery underfoot.

Types and sizes of tiles.

Three general types of ceramic tile are used on floors:

Glazed tiles are surfaced with a glasslike color glaze that is fused to the tile bodies to make them impervious to moisture and easy to clean. The glazes may be mirrorlike or semidull, mottled or textured.

As a group, glazed tiles are more suitable for installation on walls than floors, because the glazes are subject to scratching. Many kinds can be used only on walls. But if you want glazed tiles for a floor, you won't have any trouble finding some.

Standard sizes are 4¼ x 4¼ x 5/16 inches (most common), 4¼ x 6 x 5/16, and 6 x 6 x

93

5/16. Nonstandard sizes up to 1 foot square are legion. And in addition to squares and rectangles, there are octagons, hexagons, triangles, and curves.

Ceramic mosaic tiles were originally known as floor tiles because this is their principal use. Made in small units (standard sizes are 1-inch square, 2-inch square, and 1 x 2 inches) and ¼-inch thicknesses, they are mounted together on sheets measuring about 1 x 2 feet. They are extremely durable, available either glazed or unglazed.

Quarry tiles are heavy-duty floor tiles. Earthy reds or browns are the colors most associated with them, but the actual color range includes whites, yellows, greens, and blacks. Inspired by Mexican and Mediterranean tiles, American manufacturers also offer a considerable range of shapes and sizes.

Standard sizes in ½-inch thicknesses are 2¾ x 2¾, 6 x 2¾, 4 x 4, 8 x 4, and 8 x 6. In ¾-inch thicknesses, standard sizes are 8 x 2¼, 8 x 4, 6 x 6, 9 x 6, and 9 x 9.

Cost starts at roughly $1 a square foot for glazed and ceramic mosaic tiles, about ten cents less for quarry tiles. They go through the roof from there.

Estimating your needs.

Like most flooring materials, ceramic tile is sold by the square foot. So to estimate what you need for a project, you just determine the total square footage of the area to be covered and tell the dealer. If he delivers the tiles to you in unopened cartons, you will have enough extra to take care of cutting and waste, but if he breaks the cartons, you should add about 5 percent to your order.

As opposed to the field tiles, which cover all or most of the floor, trim tiles are sold by the lineal foot or piece. You probably don't need trim tiles if the floor you're laying is surrounded by walls and doors opening into other rooms. On the other hand, they are necessary if you continue the tile installation up the walls, put in a tile baseboard, or tile to the edge of a step down into an adjoining room. The following tiles are used for these purposes:

Cove-base tiles with round tops are used for baseboards. A slight upward curve at the bottom of each tile eliminates a sharp dirt-catching corner between floor and wall. Special units are available for turning right- and left-hand corners.

Coves are similar to cove-base tile but have square tops. They're used to join tile walls to tile floors. Special units are made for turning corners.

Step nosings rounded along one edge are used to edge floors above steps. (They are used for windowsills, too.)

In addition to the tiles for an installation, you must order 1 gallon of adhesive for each 50 square feet of floor and 5 pounds of dry cement grout for every 15 square feet of mosaic tiles or for every 100 square feet of larger tiles.

Preparing the subfloor.

Ceramic tiles today are generally set in adhesive. A more than adequate subfloor is formed by nailing ⅝-inch plywood to joists spaced 16 inches on center and covering with ⅜-inch exterior plywood. Be sure the plywood is securely nailed (see Chapter 2) and level. Fill voids in the top veneer with plastic wood or latex cement.

On a wood subfloor or finished wood floor you must nail down an underlayment of plywood or hardboard. If the floorboards are no more than 3 inches wide, use ¼-inch underlayment; over wider boards, use ½-inch underlayment. Lay the underlayment at right angles to the boards and space the nails around the edges and through the center 4 inches apart in both directions. Use annular-ring nails that are about three times as long as the underlayment is thick.

Although ceramic tile can be laid over resilient flooring that is firmly bonded to the subfloor, it's not a good idea, because there is no way of telling how good the bond is just by looking at it. Remove the resilient material altogether and scrape off the old adhesive.

New concrete slabs must be perfectly dry for tile adhesive to stick. The easiest way to test for dryness is to place several old rubber mats here and there around the floor. After twenty-four hours, when you take them up, if you find there is any sign of moisture on the mats or concrete, let the slab dry further.

Old concrete slabs should also be checked for dampness if you have any doubts about them. Then level them by grinding down high spots and filling low spots. Plug holes and cracks. Clean off all paint, grease, and dirt.

Laying a floor of individual tiles.

The first step in putting down a ceramic tile floor is to remove the baseboards and shoe moldings. Trim off the bottoms of door casings so tiles can be slipped underneath (see Chapter 8 for how to do this). If you're installing a threshold (the usual practice in bathrooms), order a marble threshold from the tile dealer, and cement it to the subfloor between the doorjambs with the adhesive used to lay tiles.

Getting the right arrangement of tiles in a room—the second step—takes time. Before attempting it, note how one tile abuts another. Glazed tiles generally have projecting lugs on the edges to control the spacing between them; consequently, when you place one tile against the next, you automatically make an open joint wide enough to hold cement grout. Quarry tiles, on the other hand, have square edges, so you must make the joints by eye.

These are normally ¼ to ⅜ inch wide, depending on the shape of the tiles; however, hand-crafted tiles and those made to look as if they are hand-crafted may require slightly wider joints to compensate for their rough edges.

The best way to work out the arrangement of tiles is to lay out two loose rows on the floor with the proper joint spacing. One row should run across the room, the other from end to end. Strike chalk lines on the floor perpendicular to the walls, and lay the tiles parallel with these.

In a bathroom, lay the first loose row of tiles from the threshold to the opposite wall. If the bathtub is to either side of the doorway, position the row so that the space between it and the tub is exactly divisible by the width of a single tile and joint. Then it won't be necessary to cut the tiles in the row abutting the tub. On the other hand, if the tub is opposite the doorway, the odds are that the tiles in the row next to the tub must be cut, because the distance from the threshold to the tub is rarely an exact multiple of the width of a tile. Plan the cutting so it has the least deleterious effect on the appearance of the floor. Here's how that is done:

1) After laying a loose row of tiles from the threshold to the tub, measure the unfilled space at the end of the row. If this is more than half of the width of a tile, you can't ask for a better arrangement of the tiles.

2) If the unfilled space at the end of the loose row is less than half of the width of a tile, the entire arrangement of tiles should be changed so you don't have a sliver-thick row of tiles next to the tub. To make the change, add the width of the unfilled space to the width of an uncut tile. Divide the answer by 2. Then trim the tiles in the row next to the tub and in the row next to the threshold to this figure. For example, if you're using 4¼- x 4¼-inch tiles and the unfilled space next to the tub is 1 inch, add

1 to 4¼, divide the total in half, and cut both the starting row of tiles at the doorway and the last row of tiles at the tub to a width of 2⅝ inches.

In larger rooms, if you're using fairly small tiles—say, 4¼- x 4¼-inch—start with full-width tiles along the entrance wall and one of the adjacent walls. (In other words, start in one of the front corners of the room.) Work across to the opposite walls. Thus, whatever tiles must be cut will be more or less concealed from view when you step through the doorway.

When large tiles are used in large rooms, however, the tiles at the sides of the room should be approximately equal in width. You must therefore lay the tiles from the center of the room toward the walls. For how to do this, follow the directions for laying out a resilient tile floor (Chapter 10).

If you lay tiles diagonally across a room, see directions for arranging a parquet floor (Chapter 8). In this case, no matter what the size of the tiles, they should always be laid from the center of the room.

Once you've settled on the best arrangement of tiles, you're ready to install them.

Use an organic adhesive recommended by the tile manufacturer or dealer. Spread this on the subfloor in straight strips with a clean notched trowel. Hold the trowel at about 45° to the floor and make the ridges of adhesive full, the valleys between almost bare. Cover only 2 or 3 square feet of floor at once.

Set the tiles in the adhesive with a slight twisting motion which spreads the adhesive evenly over the back, and press them down firmly. To make sure you're using the right amount of adhesive—neither too much nor too little—pull off an occasional tile and examine the underside. If it's not completely covered, you should use a little more. If it's difficult to press a tile down flush with adjacent tiles, you should use less.

After setting several tiles, level them by pressing or rubbing with a heavy wooden block, such as an 8-inch length of 2x6. If any adhesive squeezes out on the surface, scrape it out of the joints and clean the surface promptly.

When you set tiles from the middle of a wall, place several against the wall and then build out from them in a growing pyramid (see Chapter 8). When you start in a corner, set two or three tiles along each wall and fill in between, then build out from there in an ever-widening fan. When you start tiling from the center of a room, fill one-quarter of the floor at a time. In this case, the tiles are also laid in a fan shape. Don't walk on the tiles as you work. Use sheets of plywood as stepping stones until the adhesive sets.

Tiles should extend at least halfway under a wood baseboard, but should not be set closer than ¼ inch to the walls to allow for expansion. If a tiled cove is substituted for a baseboard, the cove tiles should be installed before the floor is laid.

The best way to cut ceramic tiles is to borrow or rent a tile cutter. You can do the job easily enough with an ordinary glass cutter, however. Lay a tile face up on a nonslippery surface. Wet the bottom of a yardstick with water to keep it from sliding, and hold it firmly on the tile to guide the cutter.

The cutter is held vertically between your first and second fingers, with your thumb under the handle. Press the little cutting wheel to the far side of the tile just inside the edge, and draw the cutter toward you and off the near edge of the tile in a continuous stroke.

Place the tile face up over a large finishing nail or coat-hanger wire. Center the scored line over the nail and press down on either side. This will snap the tile in two. Then smooth the edges with a file or abrasive stone.

To cut a tile to fit around a pipe or other irregular surface, trim a piece of heavy paper to the size of the tile and draw the cut on it. Then trim the paper along the line to make a template for marking the tile. Cut the tile with tile nippers. Start from an edge and work toward the cut line, nibbling out little pieces as you go along (the same way you might nibble your way across a giant cookie). Finish by shaping the cut with a file or abrasive stone.

If a hole must be cut in the center of a tile, cut the tile in two pieces through the hole, then cut out the hole in each piece with your nippers.

Laying mosaic tiles.

Mosaic tiles are easier to lay than individual tiles because they cover a lot of ground, can be arranged quickly, and are cut for the most part with scissors.

Because each tile is so small, the appearance of the floor is not seriously hurt if the units along some walls are narrower than those along the other walls. This means that you can usually start laying tiles from a front corner or front wall of the room and progress rapidly toward the opposite corner or back wall. What's more, in a small room, it's feasible to lay the sheets of tile loosely on the floor and cut the last sheets before starting actual installation.

To cut the sheets between tiles, simply bend the sheets backward slightly and snip through the backing with a pair of shears or a sharp knife. When trimming the tiles themselves, use nippers and bite off small pieces until you reach the pencil line. Don't bother to remove the backing. File the cut edges as necessary.

Lay each sheet as closely as possible into place, and slide it the rest of the way to spread the adhesive across the back. Make sure that the individual tiles in adjacent sheets are aligned and that the joints between sheets equal the width of the joints between tiles.

Press down with your hands, and go over the sheets a second time with a block of wood.

Filling joints between tiles.

Let the adhesive in which the tiles are laid set for at least twenty-four hours before grouting the joints. Then scrape out any adhesive clogging the joints and clean the entire floor with a vacuum.

Use a commercial dry-set grout available from the tile dealer. A gray color is usually preferred for floors since it doesn't show soil as readily as white. Mix the grout with water to the consistency of a thick paste. Don't mix more than you can apply in an hour, which is the time it takes for the grout to set too hard to work.

Spread the grout over a section of floor and work it into the joints with a window-washer's squeegee or rubber sponge. Go over the floor repeatedly, packing the grout into every joint. Then scrape off the excess.

After the grout sets for ten to twenty minutes and has begun to dry, tool the joints between glazed tiles and mosaics with slightly rounded edges (called cushioned edges) with the narrow end of a toothbrush handle. Draw the handle along the joints, forcing the grout into them and at the same time smoothing the surface of the grout. The joints should be concave.

Joints between quarry tiles are ordinarily flat and flush with the surface of the tiles. Sprinkle them with dry grout and wipe them smooth with a coarse rag. But if you prefer a slightly recessed joint—which makes the floor harder to mop but gives it a slight relief—omit the dry grout.

In all cases, as soon as the grout is firm, clean the floor with a sponge wrung out in *clean* water. Go over it several times. Then polish with a clean, dry cloth.

A quarry tile floor should now be covered with sheets of polyethylene film or damp burlap for several days to allow the grout to cure. You can use the same treatment for glazed and ceramic tile floors, but a simpler procedure is to spray them with a fine mist of water once a day for three days.

Laying a floor with pregrouted tile sheets.

Pregrouted sheets of tile are the newest thing in ceramic tile flooring. The tiles used are either glazed units in relatively small sizes or mosaic tiles. They are bonded to a flexible backing (like ordinary mosaic tiles) in sheets up to 2-feet square. The joints are prefilled at the factory with a tough, flexible material such as polyurethane, silicone, or polyvinyl-chloride. The edges of the sheets are pregrouted, too.

The sheets have several advantages: Because of their size, they are installed rapidly, and they are ready for traffic just as soon as the adhesive in which they are laid has set. The smooth, tough grout is highly resistant to soiling, mildew, stains, puncturing, and abrasion. It doesn't shrink or crack. Because of its flexibility, it is not affected by any movement in the floor. And it's available not just in the usual white or gray, but also in several colors to complement the tiles.

The major disadvantage of the sheets is that your choice of tiles is rather limited.

Pregrouted sheets are installed over the same type of subfloor and in almost exactly the same way as conventional tiles. One difference is that if you're using sheets of glazed tiles, the sheets should be laid from the center of the room toward the walls so the perimeter tiles are more or less the same width. Lay out the sheets as you would resilient tiles (see Chapter 10). Sheets of mosaic tiles, however, can be laid like ordinary mosaics—straight across the floor from side to side.

Use the adhesive specified by the sheet manufacturer and spread it over as much floor as you can tile in an hour. Place each sheet in exact position tight against the adjacent sheets. Don't slide it.

An hour after the sheets are laid, roll them down with a 150-pound carpet-covered roller available from the flooring dealer. Roll in both directions. In corners and other places you can't reach with the roller, place a scrap of carpet on the tiles and pound them down with a board.

Fill joints around the edges of the completed floor and around plumbing fixtures and pipes with caulking compound sold by the tile manufacturer to match the grout between tiles.

Maintaining ceramic tile floors.

Probably the principal attraction of ceramic tile is that it's so easy to care for. Frequent mopping with water or with dilute household detergent solution is usually all you need to keep a floor clean. Rubbing with a clean towel restores the luster of glazed tile.

On very dirty floors, use a strong solution of detergent, household cleanser, or commercial tile cleaner. If this does not remove all soil from the grouted joints, scrub them with chlorine bleach. If the discoloration persists, mix a bleach-containing cleanser, such as Ajax or Comet, with water to form a soup, and swab this into the joints. Let stand for about fifteen minutes, then scrub with a stiff bristle brush, rinse, and wipe dry.

To remove mildew, scrub with chlorine bleach; for the whitish film left by hard water, use vinegar.

If discoloration of the joints is a continuing problem, treat them with a silicone sealer after thorough cleaning. This won't put a permanent stop to soiling but will help.

Repairing ceramic tile floors.

Cracked and eroded joints are quickly fixed by scraping them clean and smearing in dry-set grout. If this isn't available, use ordinary portland cement without sand. If the cemented joint is not as white as the old joints, brush on white porcelain glaze like that used on kitchen appliances.

Loose tiles should be lifted out and scraped clean. Scrape the adhesive from the subfloor. Then reset the tile in fresh tile adhesive or silicone bathtub caulking. Grout the joints after twenty-four hours.

Broken tiles that are still stuck to the subfloor are chipped out with a cold chisel. Work from the crack toward the edges—never from the edges inward, because this is likely to chip adjoining tiles. Replace the tile with a new one, setting it in tile adhesive or bathtub caulking.

Before attempting violent methods to remove stains from tiles or joints, give them a hard cleaning by the methods outlined under maintenance. Applying a slurry of bleach-containing cleanser and letting it stand for thirty minutes before scrubbing is particularly effective.

Specific stain removal agents you might also use are as follows:

Chlorine bleach for inks, dyes, and Mercurochrome.

Ammonia for iodine.

Sal soda and water for greases.

Hydrogen peroxide for blood.

Detergent in hot water followed by hydrogen peroxide or chlorine bleach for coffee, tea, fruit juices, and lipstick.

14/ BRICK

The one and only brick floor with which I have had any firsthand experience was in a passageway connecting the living room of our former home with the wing we added on. The passageway actually was more than a corridor. It was a small room in which we stored books and raised plants and which also served as the wing's entrance hall from the garden. Everyone who saw it said it was delightful, and in good part the floor was responsible for the accolades. As brick floors go, there wasn't anything special about it. But it was obviously very, very practical for that particular kind of room. It was attractive and friendly-looking. And in a Cape Cod colonial house, inserted between the oak floors of the living room and the new wing, it just looked different—unusual—to some people, I'm sure, unique.

In houses of modern design, brick floors are much more common. But I doubt if they are any more at home. Brick is a warm, adaptable material which fits in almost everywhere. The only way you could go wrong with it would be to use one of those bilious brownish-yellow bricks instead of the

100

pinks and reds and attractive new colors that have been coming on the market in very recent years.

Types available.

Although this may come as a bit of a suprise, there are many differences among bricks, and you shouldn't use just any type that comes to hand as a flooring material. For example, the most widely used bricks—those going into walls—are not suitable for flooring because they do not withstand abrasion very well. Similarly, used brick—which to my mind is the most beautiful of all brick—is a poor flooring material because it has lost much of its strength, and it cracks and chips easily.

If you expect durability in a floor (and who doesn't?), the only proper type of bricks to use are those made for the purpose. These are called paving bricks or, simply, pavers. They are denser than common bricks and less absorptive; consequently, they last longer, are resistant to abrasion, and are much less likely to become stained and discolored.

Brick paving units are available in roughly forty sizes and shapes, including squares and hexagons as well as the usual rectangles. (In addition, some manufacturers sell special shapes such as bullnoses and stair treads.) Thicknesses range from ¼ to 2½ inches.

Among rectangular bricks, widths range from 3⅜ to 4 inches, lengths from 7½ to 11¾ inches. Squares run from 4 to 16 inches across. Hexagons are made in 6-, 8-, and 12-inch sizes.

Naturally, costs vary with the size and shape, but if you stick with conventional pavers, they should come to no more than about 40 cents a square foot.

Estimating your needs.

Like exterior brick pavements, brick floors can be laid with or without mortar. When laid with mortar, the bricks are separated by joints measuring ¼ inch (rare), ⅜ inch, or ½ inch wide. When laid without mortar, the bricks are jammed together as close as possible.

Before ordering bricks for a floor, you must decide (1) whether the floor will be built with or without mortar and (2) which bond, or pattern, you will use. You must also take note that bricks, like lumber, are identified by their nominal size even though their actual size is somewhat less. (For example, most bricks are said to measure 4 x 8 inches although the great majority measure less than that.)

Generally, in mortarless construction, if you want to lay bricks in any bond that suits your fancy, you must use rectangular pavers that are exactly twice as long as they are wide (in other words, they should measure precisely 4 x 8 inches). If you use bricks that are only a nominal 4 x 8 inches, your choice of bonds is limited to the running and stack bonds.

On the other hand, if you're building a floor with mortar, you must use rectangular bricks

that—though they have nominal dimensions of 4 x 8 inches—actually measure less—say, 3⅝ x 7⅝ inches. With these, you can make any type of bond, since the mortar joints allow you to line up the bricks properly. If you were to use actual 4 x 8 bricks, this would be impossible except in the running and stack bonds.

Similarly, you should use full-size square bricks in mortarless construction, fractional sizes in construction with mortar. (Hexagons are made only in full sizes.)

To determine how many bricks you need for a floor, figure the square-foot area of the floor. Then multiply by the figure given in the following table for the brick size you use:

Bricks for mortarless flooring

FACE DIMENSION OF BRICKS (actual inches)	MULTIPLY BY THIS FIGURE (bricks per square foot)
4 x 8	4.5
4 x 4	9.0
6 x 6	4.0
8 x 8	2.3
8 x 16	1.1
12 x 12	1.0
16 x 16	0.6
6 x 6 hexagon	4.6
8 x 8 hexagon	2.6
12 x 12 hexagon	1.2

Bricks for mortared flooring

3⅝ x 7⅝	4.5
3⅝ x 11⅝	3.0
7⅝ x 7⅝	2.3
7¾ x 7¾	2.3

To allow for waste and cutting, add 5 percent.

Laying a brick floor without mortar.

Unless there is some compelling reason to do otherwise, lay all floors without mortar. It saves an incredible amount of work yet yields a floor that is every bit as durable as one laid in mortar. Furthermore, the floor looks better.

The base for the bricks is usually concrete or plywood, but if you want to lay the bricks over an existing floor of wood, resilient material, ceramic tile, or something else, you can do that, too. In all cases, however, the floor must be level and fairly smooth.

As noted in Chapter 2, since the average brick is thicker than other flooring materials, the subfloor should be sunk so that the bricks and adjacent surfaces will be level. But this requirement can usually be ignored if ½-inch-thick pavers are used.

Because of the weight of the bricks, the spacing between joists should be reduced to 12 inches, or the number of joists should be doubled. But you can ignore this requirement also if using ½-inch bricks.

Cover the subfloor with two layers of 15-pound asphalt-saturated building felt. These need not be cemented down. Lay the strips at right angles to each other. Butt the joints.

Set the bricks together as tightly as possible. You can use any bond you like (several are illustrated) although the running bond and herringbone are preferable because they are more resistant to movement.

To prevent shifting, the bricks should be laid tight against all walls or adjacent flooring. This means, however, that the width and length of the floor must be exact multiples of the width and length of the bricks, and that is

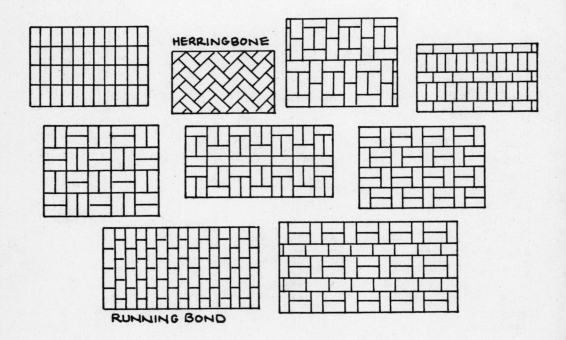

HERRINGBONE

RUNNING BOND

rarely the case. You must therefore plan some way of piecing out the flooring to fill the gaps at the edges. This is done in several ways.

If the gap at one side of the room is less than the combined thickness of a baseboard and shoe molding (approximately 1⅜ inches), simply wedge a board between the bricks and wall to fill the gap. The board will be hidden by the baseboard and molding.

If the gap at one side of the room is more than 1⅜ inches but less than 2¾ inches, wedge boards between the bricks and walls at both sides of the room.

If the gap at one side of the room is more than 2¾ inches but less than the full width of a brick, cut the bricks to fill the gap. (Score the bricks on both sides with a cold chisel, then crack them in two with a hammer and rub the cut pieces together to smooth the edges.)

When the floor is completed, spread very fine sand like that used in swimming pool filters over it and sweep it into the joints to fill irregularities and lock the bricks more tightly together. Remove the excess and give the sand about a week to work down into the joints.

Finish the floor by applying two coats of a transparent penetrating masonry sealer. This helps to protect the bricks against staining, cements the sand into the joints, and makes for easier maintenance. Use a short-napped roller or brush. The roller won't give such good penetration but is less likely to pull sand particles out of the joints. If you use a brush, take care when applying the first coat of sealer not to brush the joints too hard. If you work across them rather than lengthwise, the sand should not be dislodged.

Laying a floor in adhesive.

A new development* is ¼-inch-thick bricks, which permit you to install a real brick floor over any sound, smooth subfloor or finish floor of wood or masonry (but not resilient material). Installation is made with adhesive recommended by the manufacturer.

Lay the floor like ceramic tile (see Chapter 13). Since the bricks are made with a flange to provide joints about ⅜ inch wide, the joints must be filled after the adhesive has dried thoroughly. Use the grout sold by the manufacturer and work it into the joints.

Laying a floor in mortar.

Floors in mortar must be laid over a concrete slab made according to directions in Chapter 17. The slab should be level but need not be smooth. The bricks are then laid in a "cushion" of concrete made of 1 part masonry cement and 3 parts sand spread over the base when it is hard.

Make the cushion ½ inch thick, and spread it on the slab just ahead of laying the bricks. Until you become accustomed to the work, don't apply more than about 3 square feet of bricks at a time.

Before laying the bricks, spray them thoroughly with water or soak them in a pail so they won't draw too much moisture out of the concrete. Then embed them in the concrete cushion in any pattern you elect. Don't just set them in lightly; tap them in. As you do so, make sure the joints are of uniform width and that the bricks lie perfectly flat. Check

* Maybrik, 4545 Brazil Street, Los Angeles, Calif. 90039.

them frequently with a straight board laid across the paving. If a brick is too high, tap it down further; if too low, pick it out, spread more concrete underneath, and reset it.

Don't walk on the bricks as you work; the concrete cushion must be allowed to set. You can fill the joints any time after that, but the sooner the better, because the bricks will still retain some of the moisture applied earlier.

The joints are made in two ways. I prefer the first because, although it takes time, it produces a more reliable bond:

1) Mix 1 part masonry cement and 3 parts sand with enough water to form a grout that is just thin enough to be poured through the spout of a watering can. Pour it slowly into the joints and work it down by knifing through it with the edge of a mason's trowel. Then finish the joints with the end of a piece of ⅝-inch or ¾-inch pipe. Draw the pipe along the joint from end to end, using enough pressure to compact the concrete and give it a smooth finish. Immediately remove excess mortar from the paving and wipe off mortar stains with a damp rag. Then cure the concrete by covering the floor with damp burlap or other fabric and keeping it damp for two or three days.

2) Thoroughly blend 1 part masonry cement and 3 parts sand. Don't add water. Spread on the paving and sweep it into the joints until the joints are flush with the surface. Then with a very gentle spray of water, mist the paving until the cement-sand mix is damp from top to bottom. This is a tricky operation because if you spray the paving too much or too hard in any one place, you may dislodge the mix or give it a rough surface.

After the first spraying, let the water soak in for a couple of hours, then spray lightly again. Then spray two or three times a day for the next two days. Finally, let the mortar cure under damp burlap for a couple of days.

Whichever joint-filling method you use, the odds are that some mortar stains will be left on the bricks. Don't do anything about these for a week after the concrete is cured. Then mix 1 part muriatic acid with 9 parts water in a plastic pail and scrub the stains until they disappear. Keep the acid out of the joints as much as possible since it etches concrete. Rinse immediately with water, and rinse again. Then brush on two coats of penetrating sealer.

Maintaining brick floors.

After the penetrating sealer has dried, apply two thin coats of a water-based buffable wax, and maintain this finish as long as you can by periodic buffing. A new application of wax will be needed every six to twelve months. Try not to rewax more frequently than this, otherwise it will build up a heavy film that must be removed—and removal is difficult, because a liquid stripper clogs the joints with some of the removed soil. Scrubbing old wax with steel wool is preferable.

When a floor becomes dirty, damp-mop thoroughly with a mild solution of household detergent, and rinse. Then polish with a buffer.

Repairing brick floors.

As might be expected, brick floors escape many of the problems that afflict exterior paving. Exposed to weather, bricks themselves often chip and crack and become coated with green algae that are as slick as glass when damp. But indoors, about the worst thing that happens to bricks is that they become badly scratched or perhaps gouged. Short of replacing them, there's nothing you can do about such damage except apply penetrating sealer, wax well, and ignore it.

The most troublesome feature of brick

floors is the concrete joints. When these chip out or begin to powder and erode, scrape out the defective mortar as deeply as possible with a cold chisel or old screwdriver. Vacuum out the crumbs. Wet the edges of the bricks and pack into the joints mortar made of 1 part masonry cement and 3 parts sand. Finish the joints to match those nearby.

If the concrete in joints cracks but does not come out, simply mix a little cement with water to make a medium-thick grout and brush this into the joints.

Efflorescence—a dirty-white powdery crust—may form on brick floors laid in mortar on a slab that has not been properly protected from underground moisture. The efflorescence appears as readily on the surface of a floor finished with sealer as on one without. To remove it, scrub with 1 part muriatic acid mixed with 9 parts water, and rinse well. Repeat treatments may be necessary until the water-soluble salts in the bricks finally disappear.

Having said that bricks should be sealed to prevent staining, I should not have to outline the methods of removing stains. But just in case the seal wears off and trouble strikes, here's what to do:

Mortar stains. Scrape off the mortar with a chisel. Scrub with 1 part muriatic acid in 3 parts or more of water. Rinse well.

Paint stains. Scrape off all you can and apply paint remover. Scrub off whatever color remains with paint thinner.

Oil and grease stains. Blot up large fresh spills with paper toweling and spread cat litter over the stains for twenty-four hours. Small spills should be doused with undiluted household detergent such as Mr. Clean. Let this soak in for about fifteen minutes. Then scrub with boiling water and a brush. Rinse well.

To get rid of old stains you usually have to resort to an emulsifying agent such as Big Red or Clix, available through auto supply stores. Or make a thick paste of whiting or talc and benzine or trichloroethylene. Spread this in a thick layer over the stains and let it dry. Then brush off and repeat the treatment as necessary.

Rust stains. Scrub with a cleanser containing oxalic acid. Zud is an example.

15/ **MARBLE**

The old song, "I dreamt I dwelt in marble halls," tells pretty much the whole story about marble. It's an aristocratic building material if ever there was one. None is held in higher esteem. But it's a shame to restrict it to mansions.

If I had a yen for the kind of sumptuous bathroom that many people are building today, I'd start with a marble floor and work up. Even if my grandiose ideas about sunken tubs and enormous lavatory counters didn't work out, the room would be a smash hit just because of the floor.

I'd like marble floors in the dining room and front hall, too—not just because of their color and patterns but also because of their practicality.

And while marble isn't to my taste in my own living room, I've seen it used in such rooms with marvelous effect. I have a couple of picture booklets put out by the Marble Institute of America with marble-floored living rooms almost anyone would dream about happily. But here I go—like everyone else—straying into the mansion category again.

106

Of course, there was a time when marble was limited to big houses by its cost. But no more. Thin marble tiles are now on the market for as little as $2 a square foot.

What's more, you can build a marble tile floor yourself—just as easily as you can lay asphalt tiles.

What's available.

If you want to daydream for a second—cut and finished marble slabs are available in sizes up to about 40 square feet. Colors include black, white, blue-gray, gray, grayish-brown, grayish-pink, green (light and dark), pink, red, reddish-brown, rose, bluish-white, brownish-white, greenish-white, cream, and yellow ranging to brown.

The colors are more limited in small tiles. Even so, they range from black through grays, browns and rose to white—in both polished (gloss) and satin (honed) finishes. Sizes are a nominal 8 x 8 inches and 12 x 12 inches. All tiles are ½ inch thick.

Estimating your needs.

To order foot-square tiles, just figure the square footage of the floor and ask for the same number plus enough extras to cover possible waste in cutting.

If using 8-inch-square tiles, find the length and width of the floor in inches, divide each figure by 8, and multiply the answers.

Laying a marble tile floor.

Marble tiles can be laid on a bone-dry, grease- and paint-free concrete slab, on a conventional subfloor of ⅝-inch plywood, or on 1-inch tongue-and-groove boards. They can also be laid over an existing floor of wood or resilient tile provided a ¼-inch plywood or hardboard underlayment is installed (see Chapter 10). Whatever the subfloor, it must be level and sound. Grind down high spots in concrete. Fill low spots in all subfloors with latex cement.

Ideally, you should lay the tiles from the center of the floor toward the walls so that the border tiles are of approximately equal width. However, because of the expense of marble and the difficulty of cutting it, you may prefer to start in a corner and lay the first tiles along the adjacent walls then work across the room to the opposite walls. This does not produce a perfect installation, but if the cut tiles along the last two walls are more or less hidden from general view by furniture, the result is not objectionable.

To cut marble, use a circular saw with a fine-toothed carbide-tipped blade. You can also score marble with a glass cutter, break it in two, and smooth the edges with a file, carborundum stone, and nippers.

To bond tiles to the subfloor, use a water-resistant organic adhesive like that for ceramic tile. This is applied just ahead of your work with a notched trowel held at a 45° angle to the floor.

Lay tiles with 3/32-inch joints. Drop each unit as closely as possible into its proper position, then twist it very slightly into the adhesive. Don't walk on the floor (unless you lay plywood over it) for about twenty-four hours. Then grout the joints as soon as possible.

Use a dry-set grout like that for ceramic tiles. Spread it on the marble and work it well into the joints with a squeegee or rubber sponge. Strike the surface off flush with the square edges of the tiles with cheesecloth, and wipe up as much of the cement as possible. When the grout starts to set, go over the entire floor repeatedly with a clean, damp sponge to remove the cement film. Polish with a clean, dry cloth. Cure the joints for the next two or three days by misting with water once daily.

Finishing and maintaining a marble floor.

Despite the density of marble, an application of a transparent sealer made especially for it—for example, Vermarco Supply Company's* White Marble Seal—is recommended for maximum protection against dirt, grease, and stains. This is needed particularly on white and light-colored marble tiles.

Brush on two coats and buff both to attain a high luster. Do not wax.

---

* Vermarco Supply Company, division of Vermont Marble Company, Proctor, Vt. 05765.

If you vacuum the floor regularly to remove dust and grit, which can cause ugly scratches, maintenance of a marble floor takes a little time. Fairly frequent buffing is generally all that's needed to maintain the luster, but you should occasionally damp-mop with water or, if the floor is very dirty, with a mild detergent solution. If this doesn't work on every spot, use either Vermarco's Marble Cleaner or Wyandotte Detergent. The former is applied as a liquid, the latter as a poultice.

Apply a new coat of sealer after thorough washing with detergent about every four months. It's not necessary to reseal the entire floor if it is worn just in spots. Sealer applied to isolated areas and buffed well blends into the surrounding areas without showing lap marks.

Repairing a marble floor.

Despite application of a sealer, you should never assume that it guarantees marble against staining. The stone is rather easily discolored or etched by a number of substances used in homes. Lemon juice, vinegar, other acids, and carbonated drinks are especially injurious if spills are not wiped up immediately.

For stains that stubbornly refuse to go away after regular washing, use Vermarco's hydrogen peroxide 35%. Ordinary peroxide of the same strength is equally good. Pour the peroxide on the stain and add a few drops of household ammonia. When bubbling stops, rinse with water and repeat treatment as necessary.

On difficult oil and grease stains use Vermarco's 50-50 Liquid Cleaner and sponge on the stain for a few minutes. Then rinse. If several treatments don't correct matters, mix the cleaner with whiting or talc to form a thick paste. Spread this in a ¼-inch layer over the stains and cover with damp rags or plastic film for twenty-four hours. Scrape off the poultice and repeat if necessary.

Rust stains require Vermarco's Crystal Cleaner. Follow the instructions on the container.

Paint splatters, if allowed to harden, should be scraped off with a razor blade. Then apply paint remover. If any color remains, make a poultice of hydrogen peroxide and whiting or talc, spread on the stains, and let dry before scraping off.

Minute scratches in marble can be obliterated by rubbing with an ultrafine grade of sandpaper. Then dip a soft damp rag in powdered tin oxide available from a marble dealer and rub the area until it is shiny smooth. The same treatment can be used on small etched areas. But if etching is widespread, it's best to call in a marble finisher or dealer.

To repair breaks in grouted joints, scrape out the loose grout with an old knife. Blow out the dust. Dampen the edges with water. Then fill with white portland cement grout.

In the unlikely event that a marble tile is cracked but still stuck to the subfloor, the only solution is to chip it out completely and replace with a new tile. Take care not to damage the edges of adjacent tiles. If a cracked tile is loose, however, you can lift it out and glue it together with epoxy adhesive before resetting.

16/ SLATE AND FLAGSTONE

Not enough people have considered the possibilities of stone for flooring. This is rather surprising in view of the material's popularity for outdoor pavements. If it's considered good for a terrace, why isn't it equally good for a front hall, family room, dining room, bathroom, or kitchen?

I can answer the question at least as far as the kitchen is concerned. Stone is hard and unyielding and it gets pretty tiring to walk around on for three or four hours a day.

But what is wrong with it in those other rooms? Nothing really. On the contrary, it has a good many things that are right with it.

It's beautiful. Like that other great natural material, wood, stone has an innate beauty that is hard to define but impossible not to see.

It's durable. Fantastically durable.

It requires only minimum maintenance to keep it looking attractive.

It's impervious to moisture and most everything else that may assault it.

It's nonslippery.

It's fireproof.

And it isn't all that expensive if you don't have to have it shipped over a great distance.

You can buy cut stone tiles in ideal flooring thicknesses for roughly 65 cents a square foot. Irregular flagging sells for less than half that figure.

Slate flooring.

Slate is a very dense, fine-grained stone with horizontal and vertical grain that permits it to be split along parallel cleavage planes. It is extremely nonabsorptive, even after long immersion in water, resistant to chemicals, and easy to clean.

It is also very handsome. Colors vary with the quarry. You can get blue-black, grays ranging from blue-gray to gray-black, green, purple, mottled green-purple, and red. Some slates contain bands of darker color (called ribbon stock); others are a solid color (clear stock).

The most popular finish is the natural cleft finish, which is moderately rough with some variation in texture. Sand-rubbed finish with a slight grain of stipple in an even plane is achieved by rubbing with wet sand. Honed finish, which is rarely used in floors, is smooth and partially polished but without excessive sheen.

Most flooring slates are rectangular or square and come in sizes ranging from about 6 x 6 inches to 24 x 18 inches. Slates in irregular shapes that are fitted together like pieces in a jigsaw puzzle range from about 1½ to 4 square feet.

The standard thickness for residential floors is ½ inch, but you can just as well use ¼ or ⅜ inch. Other thicknesses up to 1 inch are available.

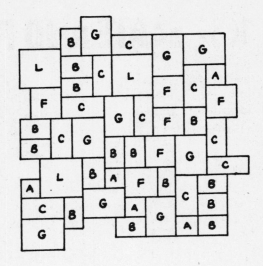

Estimating your needs.

To determine the amount of slate required, figure the exact square footage of the area to be covered. Make your measurements from wall to wall and subtract 1 inch to allow for easy fitting at each wall (an expansion joint is unnecessary). If you use irregular slates, add 10 percent for cutting and waste.

If you use rectangular and/or square slates, ask the manufacturer to give you the stock patterns that he has developed for placing slates. These not only save money and labor but also eliminate waste because you have to do little if any cutting on the job. The patterns are numbered to show exactly where each slate should be placed in the center of the floor; odd pieces around the edges of the pattern are filled in as you think best. Some of the patterns use slates of one, two, or three sizes; others use slates of numerous sizes. Selection of the pattern depends on the size of the floor as well as on your personal preference. Patterns made up of small pieces are more suited to small floors, while those with large pieces are better for larger floors.

You should also advise the manufacturer how you intend to lay the slates. If they are to be put down in mastic, they should be gauged to an even thickness.

Laying a slate floor without mortar joints.

This way of laying slate floors is gaining popularity because the work is easy, gives a uniform overall effect, and eliminates the problem—sometimes encountered with mortar joints—of concrete leaching into and staining the slates. However, the method is limited to slates of a single size laid in one of the standard bonds for bricks. Installation is made without any adhesive.

The subfloor can be of concrete, plywood, or wood. It must be sound, smooth, and level. Although a ¾-inch plywood subfloor forms a satisfactory base, it is better to use two layers of plywood such as a ½- or ⅝-inch bottom layer covered with a ⅜- or ¼-inch top layer. Hardboard can be substituted for the top layer of plywood. The sheets in the two layers should be placed at right angles to one another.

Nail ¼-inch plywood or hardboard over a board subfloor or existing finish floor.

Use ½-inch slates so they won't pop out of position if there is any play in the floor. Set the slates directly on the subfloor as close together as possible, and lock the entire floor in place with strips of wood inserted between the perimeter and the walls. Because the edges of the slates are square and smooth, there is no need to fill the joints—like brick joints—with sand.

To cut slates, use a 14-point hacksaw blade and saw with the bottom side up. A circular saw with a masonry blade can be used, too.

Laying a slate floor in adhesive.

Laying slates in adhesive produces a more rigid floor than is possible without adhesive, and it does not limit your choice of overall pattern as long as you use rectangular and square slates only. You can lay the slates edge to edge, without mortar in the joints, if you use units of a single size and follow one of the patterns for putting down bricks. Or you can use any other pattern if you provide ½-inch mortar joints.

Since the slates are cemented down, you can use a ¼- or ⅜-inch thickness without fear of their cracking. As noted before, they must be gauged so they will sit evenly in the adhesive and all will be the same height.

The subfloor should be built to the specifications described above. It should be free of paint, grease, and oil to assure that the adhesive will form a tight bond. Concrete must be totally dry.

Use an organic adhesive like that employed in setting ceramic tile (see Chapter 13). Spread the adhesive on the subfloor with a notched trowel that produces ¼-inch beads separated by clean ¼-inch valleys.

Start laying slates in a corner and work diagonally across the room to the opposite corner. If you're going to make mortar joints, set each slate by holding it at a slight angle to the subfloor, setting the low edge into the adhesive, dropping the high edge, and pressing firmly into place. If you're omitting mortar joints, simply hold the slate tight against the adjacent slates, drop it into place, and press down. Don't slide the slates any more than necessary.

Don't step on newly laid slates for about twelve hours. If you must walk across them, cover them with plywood.

Once a floor without mortar joints is completed and the adhesive has dried for three days, it is ready for service.

To fill mortar joints after the adhesive has dried, mix 1 part portland cement and 2 parts sand with enough water to form a grout just thick enough to pour through a watering can spout. Use white cement for white joints, standard cement for gray joints. Dribble this into the joints and tool it smooth, when it starts to set, with a piece of ¾-inch pipe. Wipe off stains left on the surface of the slates with a damp rag. If any traces of mortar remain after the joints have hardened, go over them again immediately with a clean wet rag. If stains persist, scrub with 1 part muriatic acid mixed with 9 parts water, and rinse well.

Laying a slate floor in concrete.

In view of the simplicity of the preceding installation methods, there isn't much reason for laying a slate floor in concrete, and the method should be used only if the subfloor is concrete. You can use slates of any thickness.

Lay the slates in a ½- to 1-inch-thick bed or cushion of mortar made of 1 part portland cement and 3 parts sand. Tamp each slate firmly into the bed and check it with a board or carpenter's level to make sure it lies flat and flush with the adjacent slates. Then lift the slate up straight—without slanting—from the mortar bed and brush the entire back with pure cement and water mixed to the consistency of a thick paste. Immediately set the slate back into position in the mortar bed and tamp it lightly.

Don't walk on the slates for several hours, until the mortar has begun to set. Then lay plywood on them and work from this as you fill the joints with a 1:2 grout mixture poured from a watering can.

Finishing and maintaining a slate floor.

You can clean a slate floor laid without adhesive or mortar immediately, but wait two weeks before cleaning other constructions. Wash with trisodium phosphate or other good general-purpose detergent, rinse thoroughly, and dry with a soft towel or chamois.

Because slate is so dense, it is impervious to most stains and does not need to be sealed. But a sealer does increase protection in areas, such as kitchens, where numerous staining agents are used.

Waxing is also optional. If you like the gloss it gives a floor, use a water-based buffable wax. Regular buffing will maintain the appearance of the floor better than too-frequent applications of new wax.

For the most part, all a slate floor needs to keep it looking well is periodic washing with detergent (Murphy's Oil Soap is particularly recommended by slate manufacturers) interspersed between regular damp-moppings.

Repairing slate floors.

If mortar joints disintegrate in a slate floor, scrape out the weak concrete and remove the remaining crumbs with a vacuum cleaner. Then fill the joints with 1 part portland cement and 2 parts sand, and strike them off to match the surrounding joints.

Hairline cracks in mortar joints are filled by brushing with a soupy grout of cement.

If a slate that was stuck down in adhesive becomes loose, lift it out and clean off the adhesive on the back and subfloor. Apply new adhesive of the same kind with a notched trowel.

If a slate is loosened from a mortar bed, remove the mortar on the back and scrape about ⅛ inch off the top of the bed. Spread adhesive into the void and stick the slate down in this. Note that in making this or the preceding repair, if you don't know what adhesive was formerly used or is recommended, you can do an excellent job by putting crisscross beads of silicone rubber adhesive (bathtub caulking) on the back of the slate.

Whitish stains appearing on slates adjacent to mortar joints are caused by leaching of the mortar. They are a rather common headache but can generally be removed by scrubbing with detergent solution. If this doesn't work, use dilute muriatic acid and rinse well.

If other stains prove troublesome, handle them like stains on bricks (see Chapter 14).

Flagstone flooring.

Strictly speaking, flagstone is not a kind of rock. It is, rather, any kind of large, thin, flat stone that is used as a paving material. The specific rocks most often meeting this definition are sandstone, limestone, and quartzite. You can, of course, order any one of these you like, but if shipping costs are prohibitive, you will probably settle for whatever is quarried and produced nearby. When an Easterner asks for flagstones, for instance, he usually winds up with sandstone flagstones. A Mid-Westerner, on the other hand, usually get limestone flagstones.

Whatever the actual stone, flagstone is a splendid flooring material. Because of its sandy-rough, skidproof surface, it has a more informal air than slate. It doesn't seem to be quite so adaptable to so many situations in the home, but there's no denying that it's good-looking, and once it's down, you can count on its lasting till the United States celebrates its quicentennial.

My favorite flagstone color is a medium-gray, but there are plenty of people who prefer something else—and there's a lot else: tan, brown, red, pink, blue, purple-green, white, buff, yellow, orange, and combinations of these. The combinations tend to be a bit gruesome. But if you want a floor that's truly hideous, mix stones of different color. Sometimes, this is done unintentionally. The homeowner orders, say, 50 square feet of flagstone from the nearest dealer and he is given a mixture of gray, blue, and tan stones. The odd thing is that when any one of these colors is seen by itself, it passes for gray and is quite acceptable. But when laid together, the grays bear no resemblance to one another. The moral is clear: When you buy flagstones, go to the source and pick them out yourself.

Like slate, flagstones come in precut squares and rectangles and in irregular pieces. Sizes vary. In the ½-inch thickness which is best for flooring, squares and rectangles are available in stock sizes ranging upward in 6-inch multiples from a nominal 12 x 12 inches (the actual measurement is 11½ x 11½ inches) to 24 x 36 inches. These sell for approximately 80 cents a square foot at the quarry. You can also buy stones cut to any size for about $1.50 a square foot.

Ordering flagstones.

If you intend to lay a flagstone floor to your own pattern, figure the square footage of the floor and add 5 percent for waste if you use square or rectangular stones, 10 percent if you use irregular stones.

The much simpler but more expensive way to order stones is to send a floor plan of the room to the flagstone manufacturer and ask him to develop a pattern for laying the stones and to supply you with all stones precisely cut and numbered to fit the pattern.

Laying a flagstone floor.

Unlike slates, flagstones are not gauged to a uniform thickness; consequently, they must be laid in a concrete mortar bed 1 inch thick. This means they are best laid on a concrete subfloor. But it's possible to lay them on a plywood subfloor provided (1) the joist thickness or spacing is increased to support the weight; (2) the subfloor is made of ⅝-inch plywood or 1-inch boards; and (3) a layer of 15-pound asphalt-saturated building felt is inserted between the plywood and concrete cushion.

Mortar for the cushion is made of 1 part portland cement and 3 parts sand. Lay the flagstones and finish the ½-inch joints between them according to the directions for laying slates in concrete: Drop each stone into the mortar bed and level it. Lift it out and coat the back with pure cement paste. Then reset the stone and firm it. Grout the joints with a watering can as soon as the cushion has set. Tool the joints with a pipe. And wash off mortar stains on the stones at once. Then cover the floor for two or three days with damp burlap.

If it's necessary to cut stones during installation, score them on both sides with a cold chisel and break them along this line with a hammer. You can also saw stones with a circular saw equipped with a masonry blade.

Finishing and maintaining flagstones.

Because flagstones are not as dense as slate and the surface is rougher, they are more readily stained and harder to clean. So after the mortar has cured for a fortnight and the final mortar stains have been removed with muriatic acid, they should be given a couple of coats of penetrating masonry sealer and then waxed with a water-based buffable wax. Thereafter, maintain them like brick flooring.

Repairing flagstone flooring.

If the mortar joints crack open or stones loosen from the base, repair like slate. But the main problem is stains—if you don't apply a sealer to minimize them. Should any appear, treat them at once by the methods given for brick (see Chapter 14).

17/ CONCRETE

Except in basements and utility rooms, finish floors are almost never made of concrete. But in new houses, concrete subfloors are prevalent.

Actually, there is no difference between a concrete subfloor and a concrete finish floor. They're built in the same way. They look the same. They have the same characteristics and raise the same problems. From here on, therefore, let's consider them all as one thing and refer to them simply as concrete floors.

I don't know whether you will ever have occasion to build a concrete floor. Most people don't. If you do, you won't find the work particularly easy. I don't mean by this that it is terribly tricky. But there are many details to attend to, and the muscular effort is tremendous. Get in tiptop physical condition beforehand, and line up at least one equally sturdy helper.

On the other hand, maintaining and repairing concrete floors is an almost universal pastime of homeowners. I have never heard anyone say it's a happy pastime, however, because you usually undertake the work only when a floor is causing troubles. That gets

you off to a bad start. And you may finish badly because the work is often frustrating.

I am not painting a very attractive picture, I know. But there's no point in trying to pull the wool over your eyes.

Concrete for floors.

Concrete for floors is a mixture of 1 sack (which equals 1 cubic foot) of portland cement, 2¼ cubic feet of clean builder's sand, and 3 cubic feet of coarse aggregate (pebbles, crushed rock, etc.) no more than ¾ inch in diameter.

The simplest way to acquire concrete to build a floor is to buy it ready-mixed and ready-to-pour, and have it delivered to the job in a cement truck. Although most suppliers make deliveries only in batches of at least 5 cubic yards, you can undoubtedly find someone who will fill smaller orders. In no case, however, can you expect the cement truck to park in your driveway for two or three or four hours while you leisurely ladle out concrete one wheelbarrow load at a time. On the con-

115

trary, you're expected to work fast, and that means you must have a helper or two to wheel the concrete from the end of the cement truck's chute to wherever you need it, to tamp the concrete down, and then to strike it off and finish it.

If you mix your own concrete in small batches, you can choose between renting a cement mixer and mixing the various ingredients with shovel and hoe on a large platform of plywood. In either case, the quality of the concrete depends on how carefully you measure out the ingredients and blend them. For measuring sand and coarse aggregate, build a box with inside measurements of 12 x 12 x 12 inches. For measuring water, use a pail with gallon markings.

Whether mixing concrete by hand or in a mixer, combine the sand and cement first. When this has been blended to an even gray color, add the coarse aggregate and blend until it is spread throughout the mixture. Then slowly add water while continuing to mix until every particle of sand and aggregate is covered with cement paste.

The amount of water used is important since the strength, durability, and watertightness of concrete are controlled by it. To determine how much to use, test the wetness of the sand (before it is mixed with cement) by pressing it together in your hand. If it crumbles when you open your hand, it is damp, in which case you should use 5½ gallons of water for each sack of cement. If the sand forms a ball that holds its shape when you open your hand, it is wet, and you should use 5 gallons of water for each sack of cement. If the sand sparkles and actually makes your hand wet, it is very wet, and you should use only 4¼ gallons of water for each sack of cement.

Estimating materials needed.

Multiply the length of the floor in feet by the width in feet and by the thickness in a fraction of a foot. This gives you the number of cubic feet of concrete in the floor. Then divide by 27 to find the number of cubic yards. If you are using ready-mixed concrete, order this amount from the supplier.

If you mix your own concrete, multiply the number of cubic yards needed by the following figures:*

6¼ to find sacks of cement required.
14 to find cubic feet of sand required.
19 to find cubic feet of coarse aggregate required.

For example, if you are constructing a 4-inch-thick concrete floor 25 feet long by 16 feet wide, multiply 25 x 16 x 1/3 to get 133 cubic feet or 4.93 cubic yards.

You must order 5 cubic yards of ready-mixed concrete.

To determine the ingredients needed for concrete you mix yourself, multiply:

4.93 x 6¼ = 31 sacks of cement.
4.93 x 14 = 69 cubic feet of sand.
4.93 x 19 = 94 cubic feet of coarse aggregate.

---

* 1 cubic yard of a 1:2¼:3 concrete mix is made with 6¼ sacks portland cement, 14 cubic feet sand, and 19 cubic feet coarse aggregate.

Preparing the site.

If the floor is to be laid within foundation walls, the latter must be constructed first. Then the ground under the slab is dug out (if it wasn't dug when the trenches for the foundations were dug) and more or less leveled. All sods, wood scraps, and other vegetable matter must be removed. If any spots are exceptionally soft, dig them out as deeply as feasible and fill with bank-run gravel. Tamp the earth throughout the floor area well. Then pour in about 6 inches of crushed stone or gravel, level it, and tamp. Lay 4-mil polyethylene film over it in a continuous sheet to serve as a moisture barrier. (If using narrow sheets, lap the edges about 6 inches.)

If you're laying a floating slab, which incorporates the foundations, follow the same procedure. You must, however, make the excavation about a foot deeper at the perimeter of the slab, and when you fill the excavation with crushed rock, it should remain a foot deeper than the rock in the center of the floor.

Building forms.

Floors are usually poured in sections 10 to 15 feet wide. The length of each section depends on how much concrete you can pour in a day. If you're constructing the slab within foundation walls and the entire slab is less than 15 feet wide and not a great deal longer, you don't need to divide the area into sections, but if the slab is wider than 15 feet, you do.

Around the edges of the slab—next to the foundation walls—the forms are made of two pieces of 4-inch-wide bevel siding. The piece next to the foundations is laid with the thin edge down; the other piece is placed against this with the thin edge up. The two pieces thus form a rectangle. Use a single masonry nail to hold each length of siding to the foundations, but don't drive the nail in tight since you must remove it before pouring concrete against it.

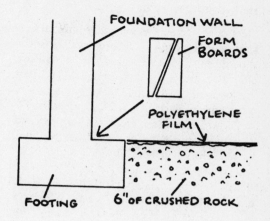

In the center of the floor, use 2x4s for forms, and hold them upright with closely spaced, sturdy stakes driven into the ground. Inasmuch as 2x4s are actually only 3½ inches wide and the slab should be a full 4 inches thick, they must be nailed to the stakes to hold them ½ inch above the moisture barrier. The tops of the stakes should not extend above the tops of the 2x4s.

Forms for a floating slab must surround the entire slab. These are made of ¾-inch plywood nailed to 2x4s driven into the ground and braced with 2x4 diagonals to keep them from spreading under the weight of the concrete. Through the center of the slab, use 2x4s nailed to stakes as described above.

All forms must, of course, be absolutely level so the floor will be level. To keep the forms from sticking to the concrete, coat them with no. 30 motor oil thinned about half and half with kerosene. All forms must be cleaned and reoiled each time they are used.

Reinforcing a slab.

Steel reinforcement is generally omitted in a slab poured within foundation walls. But it should be used in all floating slabs and in any other slabs that are laid on ground that is often wet or is not completely stable. Use a 6- x 6-inch mesh made of no. 10 gauge wires.

The strips of mesh are wired together and supported above the polyethylene moisture barrier either on special steel supports or chunks of brick or stones. The top of the mesh should be approximately 1½ inches below the top of the finished floor. (However, if the slab is divided into sections by wood forms, the mesh is placed under the forms and must therefore be sunk several inches below the top of the slab at those points.)

Don't bother to remove rust from the reinforcement. It does no harm. When pouring the slab, however, take pains to work the concrete under the mesh and around it so that it is completely embedded.

Pouring a floor.

Concrete must be poured in forms within forty-five minutes after mixing. Start pouring at a point farthest from the source of concrete and work backward. If the floor is within foundation walls, pour the first concrete in a corner against the beveled siding forms so that it will hold the forms upright. (Pull out the nail that initially held the forms when you get to it.)

Pour in the concrete slightly higher than the forms, and spade it well to eliminate voids and make it settle. Thorough spading is especially important next to the forms to produce an even, dense surface when the forms are removed.

After you've filled the forms from side to side and a few feet in from the end at which you started, strike off the concrete with a 2x4 laid across the top of the forms. Saw the 2x4

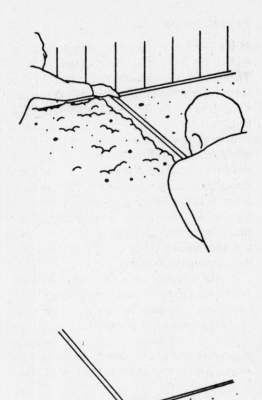

back and forth as you gradually pull it across the concrete. This further compacts the concrete and at the same time brings it down to the level of the forms. If there are any low spots in the concrete, toss in some more concrete and strike it off again.

After a strip of concrete is struck off, it should be further compacted and leveled with a bull float. This is nothing more than a board about 8 inches wide and 3 feet long with a long handle attached to the top side like the handle on a long-handled scrub brush. Place the board flat on the concrete and work it back and forth to eliminate high and low spots left by the 2x4.

Continue pouring concrete and leveling it until that section of floor is completed. If you find you can't complete the section the day you started on it, install an additional 2x4 across the section and stake it in place to hold the concrete. If you don't do this, the unsupported concrete will spread and sag; and you may not be able to bring it up to level the next day when you start pouring again, because portland cement concrete cannot be poured in layers less than ½ inch thick.

As soon as the concrete you have leveled begins to stiffen and has lost its water sheen (this may happen long before you have finished pouring operations), it should be given its final finish. The first step is to go over it with a hand tool called a float. For the smoothest surface texture, rent a metal float. You can, however, make a float out of a smooth 4-inch board about 15 inches long. Attach a handle, similar to the U-shaped steel pulls on tool shed doors, to one side in the center. To use the float, lay a piece of plywood on the concrete to kneel on. Then, holding the float flat on the concrete, sweep it back and forth and in circles until the surface is smooth and free of voids and projecting bits of aggregate.

If this work is done carefully, the finished surface will serve as a satisfactory base for other flooring materials such as resilient or ceramic tile or parquet. The surface is also suitable for use as a basement, utility room, or workshop floor. It is rough enough to be skid-proof but not too rough to look good when painted. On the other hand, it is not smooth enough for children to play on very happily—especially if their knees are bare.

Troweling immediately follows floating if you want to give the concrete a very smooth finish. The trowel used is a thin steel rectangle similar to a float or the spreader used for putting down adhesive for resilient tile floors. Working from a piece of plywood, hold the trowel perfectly flat and move it back and forth over the concrete with fairly heavy pressure. After completing the entire floor, you may go over it again for a still smoother finish.

As soon as the finished concrete has hardened enough so your finger leaves no impression, cover it with damp burlap or canvas for at least three days. Keep the coverings damp. The alternative is to cover with dry polyethylene film. The treatment is necessary to cure the concrete to maximum strength and abrasion resistance.

When the next section of a floor is to be poured, remove the forms separating it from the sections already completed. But don't remove the forms around the perimeter of the floor for several days. (The forms made with beveled siding are removed by prying out the pieces next to the foundations and then taking out the inner pieces.)

Fill the open joints between a floor and foundation walls with hot asphalt.

Finishing a concrete floor.

Finishing is optional. If desired, use epoxy paint (see Chapter 5). Let the new concrete dry for at least two weeks before application.

If you want colored concrete without a finish, a dry mineral oxide pigment should be mixed into the concrete before it is poured. (However, for white concrete, you need only white portland cement and white sand.) Pigments in various colors are available from a masonry supplies outlet; don't substitute anything else. Mix the pigment with white cement rather than the standard gray, particularly if you are aiming for a light color.

The amount of pigment used should not exceed 10 percent of the weight of the cement. In other words, for each sack of cement (weighing 94 pounds), use no more than 9.4 pounds of pigment. Mix the pigment very thoroughly into the cement and sand before adding coarse aggregate and water.

Since you can't be certain whether the concrete will turn out to be the exact color you hope for, make up one or more small trial batches several days before laying the floor. Once you have arrived at the proper color, use exactly the same proportions of pigment thereafter. The actual pouring and finishing processes are like those described.

To prevent an unpainted concrete floor from staining, apply two coats of a transparent penetrating masonry sealer two weeks or more after the floor is completed. Then apply a couple of thin coats of water-based wax.

If you don't want to use a penetrating sealer but like a waxed finish, apply two coats of a special concrete floor wax recommended by a masonry supplies dealer. This treatment does not prevent staining, however.

Maintaining a concrete floor.

Concrete floors that have not been given any sort of finish require no maintenance other than sweeping and occasional washing with water or a strong solution of household detergent. Remove efflorescence and stains as on brick (see Chapter 14). If stains defy removal, scrub with 1 part muriatic acid in 9 parts water and rinse well. Repeat as often as necessary.

If a floor is sealed and waxed, maintain it like a brick floor.

Floors finished with epoxy paint should be kept swept or vacuumed to prevent grit from scratching the paint. Wash as necessary with water or detergent solution. Touch up chipped and scratched spots as they appear. Complete repainting should not be necessary more than every three to six years, depending on how much traffic there is through the room.

Repairing concrete floors.

Repairing concrete is a somewhat unsatisfactory process, because whatever concrete you use for patching almost never matches the original floor (particularly if the concrete has been colored integrally). This may not be upsetting in a basement or other work area, but elsewhere it means you will probably have to paint the floor whether you like it or not. But except for this problem, repairs are not difficult—only tedious and sometimes frustrating.

Cracks and holes. Hairline cracks are simply brushed hard with a slurry of concrete and water. Larger cracks and holes must be opened as deeply as possible with a cold chisel. Blow out dust with a vacuum cleaner and let the concrete dry. Then fill with latex cement and trowel smooth.

If large cracks and holes occasionally leak, cut them open to at least ½-inch width and about the same depth. Cut the sides straight up and down or, if possible, bevel them so the cracks are wider at the bottom than the top. Blow out crumbs, and wet with water (if not already damp). Then pack in a mixture of 1 part portland cement and 2 parts sand.

Cracks and holes that leak almost continuously are cut open in the same way. Fill them with quick-setting hydraulic cement which comes in small packages. Mix the cement with water in very small batches and mold it in your hands until warm. Immediately cram it into the crack and hold in place until it hardens.

Leaky joints around the edges of a floor. Fill these like ordinary cracks. Use a 1:2 mix of portland cement and sand if the joints leak only occasionally; use hydraulic cement if they leak actively.

An alternative repair for joints that leak now and then is to cut them open and fill with asphalt roofing cement containing fibers. Still another method is to scratch them open as much as possible and trowel latex cement into the corner at a 45° angle. The cement should

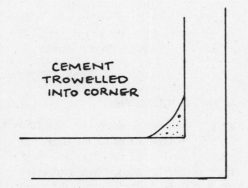

CEMENT
TROWELLED
INTO CORNER

extend at least 1 inch out over the floor and up the wall. When dry, paint the patch with a cementitious coating such as Thoroseal or with an epoxy made for waterproofing concrete.

Badly broken floors. If the floor is broken in spots, knock out the broken areas with a sledgehammer and remove the chunks. Wet the edges of the floor with water or, better, a slurry of concrete. Pour in 1 part portland cement, 2¼ parts sand, and 3 parts coarse aggregate. Spade well to eliminate voids and force the concrete firmly against the edges of the holes. Then strike off with a board and finish with a wood float. Keep the patches covered for several days with damp burlap.

If the entire floor is broken, the "easy" solution is to cover it with a new 2-inch thickness of concrete reinforced with steel mesh. A better solution is to take up the entire floor and pour a new slab.

Concrete dusting. Dusting concrete shouldn't be confused with ordinary dusty concrete. Surface dirt causes the second problem. Dusting concrete can be identified by the fact that the surface breaks down into a fine cement-colored dust more or less continuously. To put a stop to this, vacuum the floor. Then combine 1 part zinc fluosilicate with 4 parts magnesium fluosilicate. Mix ½ pound of this mixture in 1 gallon of water and mop it evenly over the floor. While it dries, mix 2 pounds of the fluosilicate mixture into 1 gallon of water and mop this on the floor when it is dry. Allow the floor to dry once more and then mop it with clear water to remove encrusted salts.

18/ **TERRAZZO**

Terrazzo is rarely seen in homes, so there's no point in devoting much space to it. Still, it's a handsome material, one of the very best from the standpoint of durability, so you just might decide to use it in your living area. What with the current interest in turning everything from barns to railroad stations into homes, you might also discover that you have acquired a terrazzo floor along with your next bank building.

True terrazzo is a mixture of two parts marble and one part portland cement. During installation, additional marble is embedded in the surface so that a finished floor surface is at least 70 percent marble. This may be of one or mixed colors.

The terrazzo is put down in large sections, which are separated by brass strips.

A newer type of terrazzo is made with marble chips embedded in an epoxy or polyester matrix. This closely resembles true terrazzo but is usually laid without divider strips. Only about ¼ inch thick and therefore a great deal lighter than true terrazzo, it can be laid over wood and plywood subfloors as well as concrete slabs. The complete installation process

122

takes only two days. Resistance to staining by chemicals is high.

Another new type of terrazzo having either a cement or plastic matrix is made in 12-inch square, hexagonal, or special-shaped tiles which are installed on any subfloor that is suitable for ceramic or resilient tile. Although not as watertight as sheet-type terrazzos, the effect of the finished floor is similar.

Laying terrazzo tile.

This is the only terrazzo you should try to install yourself. To order the tiles, find the square-foot area of the floor and order the same number of square tiles plus a few for waste. Let the dealer figure out the number of hexagonal or special-shaped tiles you need. Tiles made with a plastic matrix are ¼ or 3/16 inch thick; those with a cement matrix are ⅝ or ¾ inch. For home use, the thinner tiles are more than adequate.

Prepare the subfloor as for resilient tiles (see Chapter 10). Installation is also made in the same way, working from the center of the floor toward the edges. Place plastic tiles tightly edge to edge; leave 1/32-inch space between cement tiles and grout the joints as for ceramic tiles (see Chapter 13). Cut both kinds of tiles with a fine-toothed hacksaw blade or masonry-tipped blade on a circular saw. Plastic tiles require no finishing. Apply a marble sealer to cement tiles.

Maintaining concrete terrazzo.

Whether terrazzo is in sheet or tile form, it should be cared for in the same way.

Vacuum the floor regularly to take up grit and dirt before it is ground into the surface. Mop as necessary with water or mild detergent solution. If dirt is stubborn, use Vermarco Marble Cleaner or Wyandotte Detergent.

Although sealers are not generally recommended on true terrazzo, they may be applied to floors subjected to a great deal of traffic. As noted above, sealers should always be applied to tiles. They should be renewed—over the entire floor or only in worn areas—about every six months. Wash the floor with detergent solution first. Buff the sealer when dry.

Do not use wax on terrazzo.

To remove stains, follow directions for marble floors (see Chapter 15).

Since terrazzo is resistant to scratching and doesn't show scratches that do occur, there are few repairs to be made. If grouted joints between tiles break open, scrape them clean and refill with white portland cement grout. Broken tiles should be replaced by chipping them out (work from the break toward the edges), scraping up the old adhesive, and setting new tiles in fresh adhesive.

Large cracks, which sometimes appear in true terrazzo, should not be repaired, because the patching material looks worse than the cracks.

Maintaining and repairing plastic terrazzo.

Here again the same instructions apply to sheet and tile floors.

Sweep or vacuum the floors regularly. Wash when dirty with water or detergent solution. Don't wax.

Badly damaged sheet floors should be repaired by a professional. Replace tiles yourself.

19/ CARPET

Of all the industries I know, the carpet people go farthest in trying to keep the public in the dark—if not actually to confuse it—about its products. Manufacturers offer almost no useful information about basic carpet values and how to differentiate between them. If you write to the industry association, the Carpet Institute, you never get an answer. This leaves the homeowner in the doubtful position of having to learn what he needs to know from the dealers who are eager to sell him a $1000 broadloom, and if you think the average rug salesman knows very much about what he's selling, you're living in a dream world.

Yet for all its exaggerated statements and public-be-damned attitude, the carpet industry turns out a product that is, at its better or best, a superb flooring material. It has these advantages; it is:

Beautiful.
Extremely pleasant to walk on.
A fine insulator against a cold or hot subfloor.
An outstanding sound-deadener that not only absorbs much of the noise

reverberating within a room but also muffles the sound of footsteps.

In addition, a high-grade carpet costing $15 a square yard or more can be expected to last fifteen to twenty years without requiring major repairs or extensive maintenance (but once it is worn out, it's through).

Carpet's position among flooring materials has changed somewhat in the past thirty years. Originally, it was only a floor covering—an extra something you put down on a finish floor of wood, tiles, and so on to protect it, beautify it, and make it warmer and quieter. Today carpet is considered also as a finish floor in its own right—a substitute for wood, tile, and so on.

Credit the idea of laying carpet wall to wall for this development. If you use carpet in this way, there is no longer any real need for putting a finish floor under it. You can instead lay the carpet (with a rug cushion underneath) directly on a concrete slab or on a ½- or ⅝-inch plywood subfloor. The result is a considerable initial saving of money.

Let's say you're building a 15- x 15-foot

addition for a new family room and you're wondering whether to put down a floor of oak covered with a 12- x 12-foot rug or wall-to-wall carpet over a plywood subfloor. The cost of the two installations works out as follows:

	FINISH OAK FLOOR WITH RUG	WALL-TO-WALL CARPET
Plywood subfloor*	$46	$85
15-pound building felt	5	—
Oak floor	160	—
Finishing oak floor	30	—
Carpet and pad	192	300
Carpet installation accessories	—	15
TOTAL	$533	$400

Thus, you save more than $130 by carpeting wall to wall. Over a period of twenty to thirty years, however, the finish oak floor will probably prove to be less expensive because you'll spend roughly $100 when you replace the carpets, and you'll probably get somewhat longer life out of the 12- x 12-foot rug because, as it shows wear in some places, you can turn it around to equalize the wear over the entire surface.

A more serious (because it may be more immediate) drawback of wall-to-wall carpet is that it usually goes with the house when you sell, whereas the rug can be taken with you.

Wall-to-wall carpet versus rugs.

Despite its attractions as a "permanent" floor finish, the principal use of carpet continues to be for covering other finish floors. Here, again, wall-to-wall carpet enters the picture, and you often find yourself choosing between it and a smaller area rug. It's not an easy choice.

From the standpoint of cost, wall-to-wall carpet is at a definite disadvantage if the floor you're covering is in good condition. On the other hand, if the floor needs a major overhaul, you may find upon digging into the matter that wall-to-wall carpet comes out slightly ahead.

From the standpoint of routine day-to-day maintenance, wall-to-wall carpet is always a little better than a rug simply because it takes fewer tools and less time to clean one large expanse of a single material than a similar expanse of two different materials. Furthermore, you don't have to go to the trouble of moving furniture, turning back the rug, and waxing or polishing the wood around it.

Counterbalancing this advantage, however, is the fact that, even when laid over a finish floor, wall-to-wall carpet may still be considered part of the house that must be left behind when you move.

* The difference in cost stems from the fact that under a wood floor you need only a rough structural grade of plywood whereas under carpet you need plywood with a smooth C-plugged or better face.

Types of carpet.

Most people who buy carpets look for the best quality they can afford. But when it comes to judging quality, they're up a tree and usually depend on the salesman's opinion.

Here's how you can protect yourself against the slick talker:

Regardless of the fibers of which a carpet is made, regardless of the way it is woven, the density of the pile is the best indicator of carpet quality. To determine density accurately, you must know the weight of the fibers per square yard and the height of the pile. The salesman should have manufacturer's literature giving both figures. Compare the figures for two or three carpets. Just because they all have the same fiber weight doesn't mean they're of equal quality. If the pile of one carpet is 1 inch high whereas that of another is ½ inch high, the latter is the denser—and better—because it obviously contains more fibers. If the weights are different and the piles of the same height, the heavier sample is the denser—and better. On the other hand, if both the weights and pile heights are different, you must resort to mathematical proportioning to determine which is the denser.

Another but much less accurate method of checking carpet density is to bend a corner of the carpet backward and examine the backing. The less of this that you see, the denser the pile.

The amount of traffic a carpet will take depends on the density and also on the fiber content, which is described on the label. Nine fibers are currently in general use. Some carpets contain just one of these; others contain a blend. The way a blended carpet wears, soils, and so on, reflects the characteristics of the principal fiber in the blend.

Wool has been a carpet fiber since the days of the nomads, and it is still so highly regarded that all other fibers are compared against it. As long as a wool carpet is mothproofed (as most now are), you can't ask for much better. It is very resilient; it is crush-, abrasion-, and soil-resistant; it is durable and easy to clean. But because the wool must be imported and because competition from man-made fibers gets increasingly stiff, wool carpets have risen in price and become rather scarce.

Acrylic is more like wool than any other synthetic fiber, and carpets made of it are, accordingly, excellent. They are resilient, durable, resistant to abrasion, crushing, and soiling, and easy to clean. And like other synthetics, moths leave them alone. We've had an acrylic carpet in our much-used living room for ten years and, except for the fact that it shed so much when brand new that it clogged the vacuum cleaner, it's never been a problem. I expect it to last quite a while.

Acrilan, Creslan, Orlon, Zefran, and Zefkrome are well-known trade names for acrylic fibers.

Modacrylic is like acrylic but less resilient. It also pills (forms fuzz balls) more readily. It is frequently blended with acrylic fibers to increase the fire resistance of the carpet. Dynel and Verel are trade names.

Nylon has exceptional durability and resistance to abrasion. Its resiliency is comparable to that of acrylic and wool, and its crush resistance is also good. But unless treated, it is subject to static electricity, and it shows soil pretty badly (though it is easily cleaned). The best nylon carpets are made of textured filament nylon. Trade names include Anso, Antron, Caprolan, Celanese Nylon '66, Cumuloft, DuPont 501, Enkaloft, Nyloft, and Tycora.

Polyester is very resistant to crushing, washes well, and has good durability. But stains are hard to remove and resiliency is not very high. Avlin, Dacron, Encron, Fortrel, Kodel, and Trevira are some of the trade names.

Polypropylene, often called olefin, is one of the easiest carpet fibers to clean when it gets dirty (which it does rather reluctantly). It's also very durable, has good abrasion resistance, and is fairly resilient. But it crushes easily and lacks the fine feel and appearance of other synthetics. Trade names include Herculon, Marvess, Polycrest, and Vectra.

Rayon goes into low-cost scatter and area rugs, but these are a poor buy. Unless you're willing to put out extra money for the densest available rugs, you can expect poor resistance to soiling, poor resistance to crushing and abrasion, poor resiliency, and poor flame resistance. Even at best, rayon rugs should not be compared with most others. Trade names include Avicron, Avisco, Corval, Fibro, Kolorbon, Skybloom, and Skyloft.

Saran is very durable, cleanable, and resistant to dirt and stains, but its resistance to crushing and abrasion are only fair to good. Carpets made of the material tend to turn a little darker in color over a period of time.

Cotton ranks high in durability and washability; it's very soft and comfortable underfoot, and fairly economical. But it soils quickly and crushes badly except in unusually dense construction.

Kitchen carpet, indoor-outdoor carpet, and broadloom are made of the same fibers used in conventional carpets. The first two are usually made of polypropylene, nylon, or acrylic (the best of the fibers for the purpose). Broadloom can be made of any fiber. It gets its name from the fact that it is made on a loom more than 3 feet wide.

Carpet construction is not necessarily a guide to carpet quality. Even so, many people worry a lot about the backing on carpets. This is made of jute or man-made material, and on tufted carpets and sometimes on others the fibers are covered with a secondary backing of rubber, polypropylene, or jute to hold the yarns securely. Impressive as this may be, however, it's worth noting that Oriental rugs have very simple backings, yet they are more durable than all other carpets and also more beautiful.

But carpet construction does have a decided bearing on the appearance of carpets. Four methods are used today for machine-made carpet, several others for hand-made rugs.

In the past, all machine-made carpets were woven on big looms. The principal weaves were Axminster, velvet, Wilton and chenille. Of these, the first two were the least costly and therefore most common; the last two were generally used only in carpets of superior quality.

But times have changed. Today the majority of carpets are tufted. The yarns are punched through the fabric backing with needles, and the resulting loops are then cut or uncut. Tufted carpets are classified according to their surface texture by such semidescriptive names as loop, plush, shag, splush, frieze, twist, sculptured.

Two even newer construction methods used in carpet factories are needlepunching and flocking. In needlepunched carpet, which is most often made for indoor-outdoor service, webs of loose fiber are needled into the backing to produce a flat texture that can be embossed. In flocked carpets, also made primarily for indoor-outdoor use and kitchens, chopped fibers are simply stuck to the backing. The result is a short pile with velvety texture.

Hand-made rugs are woven and tufted and also braided, hooked, and needlepointed. Braided rugs are made by twisting fibers or strips of fabric into thick braids which are then sewn together in concentric loops. Hooked rugs are made by pulling loops of yarn or cloth strips through the backing with a hooked instrument. Needlepoint rugs, which are extremely fine and rare, are made by needling long strands of yarn through a backing in fine, tight stitches.

Laying carpet.

Carpets can be laid over any floor or subfloor that is clean and free of moisture (polypropylene carpet, however, is not harmed by moisture). The surface need not be smooth or level, but it should be reasonably so if you want the carpet to look its best and feel right underfoot.

Installation of a rug cushion under the carpet—though not mandatory—is highly advisable because it lengthens the life of the carpet, adds resiliency, muffles impact sounds, and helps to insulate against heat and cold.

Some carpets have a built-in foam or plastic backing which takes the place of a separate rug cushion, but these are rarely of top quality. So you usually must buy a cushion separate from the carpet.

The best rug cushions are made of jute fibers and hair with a rubber coating on top and bottom. They have good resilience and a high noise-reduction coefficient and are splendid insulators. Unlike the cheaper, all-hair pads, they don't ravel, and they prevent moisture on the carpet from penetrating through to the floor beneath. But they should not be used in rooms with high humidity because they may develop an unpleasant odor.

The minimum weight of a cushion should be 40 ounces per square yard.

Flat latex-foam rubber cushions up to ⅜ inch thick and with a weight of at least 56 ounces per square yard are also excellent—particularly in high-humidity areas and over radiant-heated floors. Urethane foam pads are comparable but should not be laid on heated floors because of their good insulating characteristics.

Sponge-rubber pads are used where you want extreme resiliency. But rugs tend to move around on them.

When cutting rug cushions (you need a sharp utility knife for hair pads), make them about 2 inches smaller than the rugs. Piece together small pads if you don't have a large one that will do. It's best to stick the pieces together with wide adhesive tape.

Before rolling out a rug on a cushion, position it properly at one end of the pad and then roll it out slowly. If it isn't squarely centered on the pad, don't try to pull it straight, because that will move the pad as well. Lift one side of the rug and lower it rapidly. This traps air under the rug, and as the bubble moves toward the far edge of the rug, it floats the rug so you can guide it into place.

Putting down wall-to-wall carpet.

You can save $3 or more a square yard if you put down wall-to-wall carpet yourself. It's not as difficult as you may have thought, provided you rent the necessary tools from the carpet dealer. These include a knee kicker for stretching the carpet, a stapler for 9/16-inch staples, and (possibly) a seaming iron. The only accessories needed are thermoplastic carpet tapes to join strips of carpet and tack-less strips with hundreds of tiny needlelike pins for gripping the carpet edges. One type of strip is made for wood floors, another for concrete.

Before starting the installation, fill holes in the floor and level high spots. If there are large depressions, fill them with latex cement and feather the edges (the cement will stick to everything except resilient flooring). Make sure floorboards are nailed down tight and all squeaks are silenced, because you won't be inclined to rip up the carpet and fix them later if they become annoying.

Although it's not essential, it's good to remove shoe moldings, which may otherwise become chipped and need refinishing (not an easy task when they're partly shrouded by carpet fibers). Butt the carpet tight against the baseboards.

Cut the tackless strips to fit around the edges of the room, posts, and built-ins in the center of the room. Use a large pair of tin snips or a small fine-toothed saw. Then nail the strips to the floor at the foot of the baseboards.

Roll out the rug cushion on the floor and trim it to fit within—not over—the tackless strips. Butt edges of adjacent strips neatly. Then, working from one end toward the other so you don't create bubbles, staple it to the floor along all edges. Space the staples 9 to 12 inches apart.

In a large room where you must use two or more strips of carpet, figure the layout carefully so you don't waste any more material than necessary.

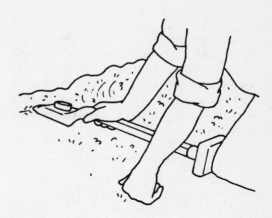

Measure the room and make a paper plan to guide you in fitting the carpet. Allow 1 inch extra at each wall to compensate for the fact that they may not be straight or at right angles to one another. Then go around the room again double-checking all measurements. Roll the carpet out on the floor (carry it into a larger room if you don't have space to maneuver it easily) and transfer the measurements to it with white chalk. Double-check them once more. Then place a piece of plywood under the carpet and cut, with carpet face up, with a sharp utility knife.

Place the carpet in position on the floor and use the knee kicker to stretch it tight and smooth. The kicker has an adjustable dial which controls the length of its teeth. Set the dial so the teeth don't dig into the carpet backing. Long-pile carpet takes a high number; short-pile uses a low number. To use the kicker, kneel on the carpet facing the wall, hold the tool by the neck, and place the toothed head flat on the carpet an inch or so back from the edge. Then hit the padded end with your knee to push the carpet on to the tackless strip. Fasten the carpet to the strip along one wall, then along one of the adjacent walls, and finally along the opposite walls.

After going all around the carpet once, return to the first two walls and repeat the process as necessary until the carpet lies perfectly flat. But be careful not to pull it out of shape.

Because carpet has a certain amount of stretch, the kicking-fastening operation will stretch it and push the edges well up on the baseboards. So you must then go around the room and trim the carpet with your knife to fit neatly against the baseboards.

At doorways and other openings, stick the carpet edge down with double-faced adhesive tape stuck to the floor underneath. The alternative is to nail it with carpet tacks. To protect an edge against scuffing, you may nail along it an L-shaped edger strip similar to that used for resilient flooring (Chapter 10).

If you have to join pieces of carpet, trim them straight so the edges butt. Center thermoplastic carpet tape under the edge of one piece. Align the edge of the other piece over the tape. Open the seam slightly and run a heated seaming iron slowly along the tape. Press the carpet edges together onto the tape as you proceed.

Inexpensive carpet is installed in much the same way. But cutting is usually done with shears, and the edges of the carpet are fastened down with double-faced adhesive tape which is first stuck to the floor. The carpet is fitted into place by hand—not with a kicker—without too much stretching. Pieces of carpet are also joined with double-faced tape.

Carpet tiles and strips.

If you don't feel up to tackling a wall-to-wall carpet installation but want the same effect, carpet tiles and strips are the answer. Tiles come in 9-, 12-, and 18-inch squares; strips come in 3- x 5-foot rectangles. They are made of polypropylene or nylon with an adhesive backing. Thus they can be glued to any smooth, clean, dry floor—even resilient flooring or waxed wood. Bare wood, plywood, and particleboard, however, should first be given a sealer coat of shellac or varnish.

Because the pile pretty well hides the seams, and the tiles (and strips) usually have an overall pattern without strong features, there is usually nothing to be gained by installing the tiles—like resilient tiles—from the center of the room toward the sides. Simply start the installation in a corner and work across the room to the opposite walls.

To lay the tiles, peel off the protective paper backing and press them down. They will hold securely without rolling or heavy pressure, yet they can be taken up at any time and stuck down again elsewhere. Place the tiles with the arrows on the back pointing in the same direction.

Maintaining carpet.

From the cleaning standpoint, the main advantage carpet has over other flooring materials is that its thick, porous surface hides much of the dirt that gets into it. But the fact that a carpet looks fairly respectable is no excuse for not taking care of it faithfully. If you let too much dirt—especially gritty dirt—accumulate, the carpet won't last very long. And while I dislike making any fearsome statements about the health dangers that lurk in dirty carpet, I think it's interesting that just the other day a major hospital in my area tore out all the carpet it had put in because it was so difficult to keep clean and—I presume—sanitary.

For long life, almost daily cleaning in heavy traffic areas is essential. You don't need to drag out a heavy-duty vacuum cleaner, but do go over the areas with a carpet sweeper or lightweight vacuum. Then at least once a week you should vacuum the entire carpet—preferably with an upright machine or tank with a beater attachment. (Suction alone doesn't lift out all the grit.)

When a carpet becomes soiled, a thorough shampooing is in order. If money is no object, call in a professional rug cleaner. Whether he removes the carpet to his shop or cleans it right there in your house, he'll do a better job than any housewife can dream of doing. Several methods are employed. My wife and I have been most satisfied with a steam extraction process.

Do-it-yourself shampooing is done either with a powder, liquid, or aerosol spray. None is as effective as it's said to be.

Powdered shampoos are used mainly for light cleaning. They're particularly recommended for wool carpets because they don't mat the fibers as liquids may, but they are equally good on other carpets. Vacuum the carpet first. Then sprinkle a section at a time with the shampoo and work it in with a soft brush or special applicator. When the entire carpet has been treated, let the shampoo "cook" for an hour. Then vacuum thoroughly.

Liquid shampoos are diluted with water according to the maker's directions, whipped into a foam, and then applied as a foam. For best results, use an electric rug shampooer or combination floor polisher and rug shampooer. If you don't own your own, you can rent one. But the foam can also be applied with a soft brush or even a sponge.

Vacuum the carpet first. Slip pieces of aluminum foil or plastic film under all furniture legs. Then apply the shampoo in even, overlapping strokes and work it in with your shampooer or brush. Finish each section of carpet by stroking in the same direction. Don't walk on the carpet until the shampoo has dried completely. This will take several hours—perhaps overnight—but you can hasten the process with a fan. When the carpet is finally dry, vacuum it.

Aerosol shampoos, like powdered types, are ready for instant use, but you pay extra for their convenience. After vacuuming the carpet and putting shields under furniture legs, spray the shampoo on the floor in overlapping strokes. Be careful not to overwet any area. Work the shampoo into the carpet with a brush or shampooer, always finishing each section with strokes in one direction. After the shampoo has dried, go over the carpet with a vacuum.

Small washable rugs are most easily shampooed in a washing machine. For a cotton rug, wash in hot or warm water, depending on how color-fast the dye is, for ten minutes. Rinse in warm water. Spin-dry, and dry in a dryer or on a line. For acrylic rugs, wash for five minutes in warm water. Rinse in cold water. After spinning, dry in an automatic dryer at low heat for about fifteen minutes. Remove the rugs while still damp, shake well, and spread out flat so they will dry without matting.

Carpet tiles are cleaned like conventional carpet. Moisture seeping down through the seams should not loosen the adhesive; even so, it's a good idea not to throw caution to the wind and wet the tiles too much.

Removing stains.

There's no need for panic when something spills on a carpet. The odds are that you'll be able to remove it without leaving a permanent spot. Even so, you should not approach the removal job nonchalantly. Get on with it just as soon as you notice the spill, because some things that may appear innocent have a nasty way of causing discoloration at a later date.

Immediacy, in other words, is the number one rule in coping with carpet stains.

The second rule is: Use a gentle, sensible removal technique. In short, don't scrub spots violently, thus making them worse. Lift off solid materials neatly with a knife or spatula. Vacuum off dry materials. *Blot* up liquids with paper towels or clean cloths.

After taking up everything possible, dampen a soft, clean, white rag or paper towel with whatever cleaning agent you decide to use and apply it to the stain with a sort of blotting—rather than an all-out scrubbing—motion. Always work from the edges of the stain toward the center so you don't spread it. Don't wet the rag too much. Turn it to a clean section as it gets dirty. Don't pour cleaning agent directly on the carpet. Finally, when the stain is gone, blot up the excess moisture with paper toweling and weight additional toweling down on top of the stain.

Rule number three is to find a couple of cleaning agents that work for you, and stick with them. Two are generally recommended by carpet experts and home economists: trichloroethylene to remove greasy stains and a homemade mixture of detergent and vinegar to remove other stains. The latter consists of 1 teaspoon of any powdered neutral detergent like that used to wash fine fabrics, 1 tablespoon of white vinegar, and 1 quart of warm water.

Directions for removing specific stains follow:

Oily foods and materials such as cream, olive oil, face cream. Scrape or blot up excess. Clean with trichloroethylene and blot dry; use trichloroethylene again if necessary. Blot and let dry; brush to restore the pile.

Sugary and starchy substances such as wine, alcoholic beverages, candy, fruit juices. Blot or scrape up excess. Clean with vinegar-detergent solution and blot dry; clean again if necessary. Brush pile when thoroughly dry.

Protein-based substances such as coffee, milk, tea, chocolate, blood, eggs. Blot or scrape up excess. Clean with vinegar-detergent solution; blot and let carpet dry. Then clean with trichloroethylene; blot, dry, and brush.

Lipstick, chewing gum, grease, tar, crayon. Remove excess. Clean with trichloroethylene and let dry. Clean with vinegar-detergent solution and dry. Use trichloroethylene again if necessary. Dry and brush carpet.

Urine. Blot thoroughly at once. Apply vinegar-detergent solution or prepared puppy stain remover. Carbonated water is also good. Blot; let dry, and brush pile.

Vomit, excrement. Waste no time in going to work on these. Scrape up as much as possible. Clean with vinegar-detergent solution and blot. Then clean with 1 tablespoon ammonia in ¾ cup water. Blot. Then rinse with 1 part white vinegar and 2 parts water. Blot and allow to dry. Brush pile.

Rust. Use a special rust-remover such as Whink, available at hardware stores.

Ballpoint pen ink. Some inks are water-soluble, but don't waste your time trying to find out whether your troublemaker falls into this category. A special ballpoint ink remover called Inknix (available at stationery stores) does a good job on all kinds of ink.

Unknown stains. Remove excess. Clean with vinegar-detergent solution, blot and dry thoroughly. If discoloration remains, apply trichloroethylene; blot and dry. Consult a professional rug cleaner if you still haven't licked the problem.

Repairing carpets.

Pills. These little fiber balls forming on the surface of some kinds of carpet are annoying but nothing to get excited about. If the vacuum cleaner doesn't scoop them up, do it with your fingers.

Sprouts. These are loose ends of pile that protrude above the carpet surface. Don't pull them out. Just trim off with shears.

Crushed spots. Dents in carpet can be minimized but not prevented by using coasters under narrow furniture legs. To restore dents, hold a steaming steam iron just above them and moisten the pile. Don't touch the iron to the carpet. Then brush the area.

Burns. If the fibers are not charred too deeply, snip off the tops and apply a vinegar-detergent solution to remove the brown color remaining. If the hole is deep, clip the burned fibers out completely. Pull clean fibers from a scrap of carpet; thread these into a needle and push them through the carpet backing into the hole. When the hole is filled, glue the ends to the rug backing with rubber cement or white glue. Then trim off the fibers flush with the carpet surface.

Frayed edges. Whip-stitch with heavy thread or cover with gummed carpet binding.

Scalloped edges. You may be able to make them lie flat by steaming with a steam iron, but don't count on it. The only real solution is to stick the edges to the floor with double-faced adhesive tape.

Badly worn areas. When you can see the carpet backing, start thinking about buying a new carpet. But if you're feeling hard pressed, you can extend the life of the old carpet to some extent by painting the worn area with a dye solution matching the surrounding area. Repaint whenever necessary.

20/ MISCELLANEOUS FLOOR COVERINGS

Recognizing the ingenuity of man in developing new materials and finding new uses for old, we can never rule out the possibility that someone will come along with a new kind of floor covering that propels carpet into limbo. Numerous carpet substitutes already exist, in fact, but for one reason or another, they have never made much headway in the United States. This is not to say, however, that they do not have their place in the home.

Rugs made of plant fibers (other than cotton).

Rugs that are usually known as grass or straw rugs even though they're made of something else are generally restricted to porches and other outdoor locations, but they lend themselves nicely to all rooms with a casual look. Primarily of coarse texture and light to medium brown color, they blend beautifully with wood-paneled walls and ceilings and floors made of wood, quarry tile, brick, flagstone, and concrete. Their cost is low, wear-resistance surprisingly high, mainte-

nance easy. Rug cushions are unnecessary, although it often pays to put several layers of newpaper under the thinner rugs to give them a little more resilience.

On the other hand, grass or straw rugs have a tendency to skid unless anchored with double-faced adhesive tape placed around the edges. And since they are usually made not in large woven pieces but of squares and strips sewn together, they do not give the overall effect of carpet.

They are made of several materials:

Grasses of various kinds. Some are almost silky; others are coarse. The rug's texture varies accordingly. For example, China grass is quite fine and therefore used in smaller weaves. Woven into long strips up to a foot wide, it makes a very pliable rug with little durability.

Sea-grass rugs, on the other hand, take very hard wear because the grass is naturally tough and the rugs much thicker. The grass is woven into lightweight rugs of up to about 30 square feet and heavyweight rugs up to 108 square feet.

Rice straw is also very durable because the

straws are sewn together into inch-wide strips which are then plaited together in an over-and-under pattern. The rugs, available in standard sizes, are occasionally treated with plastic to improve wear and soil resistance. (Japanese tatami are made of rice straw.)

Hemp. This is a tough vegetable fiber making a rug that is as nearly indestructible as hemp rope. It is produced in squares which are sewn together into rugs with a slightly furry look rather like that of a carpet.

Sisal. Another tough vegetable fiber, sisal is woven into small rugs as durable as those of hemp but with unusual softness. Unlike the other so-called grass rugs, sisal rugs take and hold dye well and are therefore available not only in browns but also in green and blue. They are often woven into intricate designs.

Rush. Rush is a marsh grass that is braided and woven into squares which are sewn into rugs. Thin rugs are short-lived, but thick rugs made of tightly twisted strands last as long as the average carpet.

Cocoa. Cocoa fibers are tougher than sisal but do not take dye and are coarse in texture. They are woven into 27-inch-wide strips which are generally used as runners (and diving board covers).

Lauhala. This Hawaiian material is made from tree leaves folded into 1-inch ribbons which are then plaited into floor strips and standard-size rugs. The rugs have a slightly glossy finish unlike that of other grass rugs, but they tend to become brittle unless sprinkled with water now and then.

All rugs made of plant fibers should be vacuumed or shaken frequently. Pick them up and vacuum the dirt that seeps through to the floor. When dirty, you can take them outdoors and hose them down or sponge them with mild detergent solution and rinse well. Dry quickly in an airy place, because they are prone to mildewing.

If the fibers in a rug are damaged, glue them together as soon as possible—before they split and ravel—with white wood glue.

Restitch torn seams with strong thread. Cut out and replace badly damaged squares or strips.

Canvas floor cloths.

In the eighteenth century it took so long for Americans to import rugs from the Orient that they started making their own out of canvas, which was given several coats of linseed oil paint and then finished in a bold, bright, usually geometrical pattern. Later, when carpet imports improved, the floor cloths disappeared, but now they are being made again in limited quantities. Decoratively, they are suited to any style of architecture and not-too-formal decoration. They wear well, especially if laid over a thin rubber pad, and when the paint starts to look thin, they are easily touched up or completely repainted.

To protect the cloths against soiling and stains, they should be treated occasionally with solvent-based paste or liquid wax.

Fur rugs.

If they're successful, big game hunters can pretty well keep their floors covered just through their own efforts. And it's a certainty that few rugs evoke more admiration than those brought home via a taxidermist.

But unfortunately, once you put down a fur rug, you must give it a lot of care. For one thing, you should keep it away from too much heat or sunlight, either of which can dry out the leather and cause it to disintegrate. In addition, sunlight may cause the fur to fade.

Vacuum the rug frequently to remove dust and embedded dirt and grit. Use suction only—not beaters—and not too much of that. Or you can simply flop the rug over and give it a gentle shake.

When the rug is soiled, send it to a fur cleaner for thorough cleaning and general revitalization. Have it mothproofed at the same time.

Rotovinyl temporary floor coverings.

In times past, linoleum was often cut into rugs. Today, some of the more flexible, colorful rotovinyls are used in the same way. These are available not only in standard rug sizes such as 9 x 12 and 12 x 15 feet, but also in continuous rolls which you can cut and lay wall to wall without any adhesive.

One problem with all such flooring is that, because of its thinness and satin-smooth surface, it eventually conforms to and reveals the surface of the floor on which it's laid. So even though you don't expect to keep it for years, you should make sure the floor is very smooth and level.

To install rotovinyl from wall to wall, follow the directions in Chapter 11 for cutting it to size and shape. Then simply roll it out on the floor and give it a couple of hours to work out the curls and lie flat.

Maintain the covering like a permanently installed resilient sheet floor.

Matting.

Although matting is really made for commercial and public buildings, it is good for covering concrete slabs in homes if your main interest is in finding something that is at once durable, pleasant underfoot, and reasonably attractive. It is also ideal for entries and halls through which people with dirty, wet feet are forever trooping.

Matting is generally made of rubber or of shallow-pile nylon laminated to a vinyl backing. It comes in long rolls in widths up to 6 feet. Easily cut with a knife or shears, strips are laid side to side to cover an entire floor.

One of the best ways to use matting—especially that with nylon pile—is to recess it in the floor just inside your exterior doors where it will automatically wipe off the shoes of everyone entering. To make an installation, when you're putting down a new hall floor, simply mark off the space for the mat and lay the floor (wood, tile, or whatever) around it. Then drop the matting—without fastening—into the recess. The top should be approximately level with the surrounding floor. If it's too low, nail plywood or hardboard into the recess under it.

Duckboards.

Duckboards are made of thick hardwood slats joined together side by side with ½-inch spaces between. One kind of duckboard is rigid, another can be rolled up like a carpet runner. They are available in standard sizes up to about 6 feet long and 5 feet wide. You can easily make your own rigid duckboard by nailing strips of wood to a couple of 1- x 3-inch cleats.

I can't think of any uses for duckboards except in work areas on concrete slabs, which are often wet or cold, but nothing else can equal them there.

Cork carpet.

Cork carpet is the ideal material for people who like to sneak around the house without making any noise. I don't know of any homeowner who has it, but the post office department uses it in hidden galleries where guards prowl looking for miscreants who may be tampering with the mails.

Generally used for making bulletin boards, the carpet is made of ground-up compressed cork particles bonded to a burlap backing. It's produced in rolls up to 90 feet long, 6 feet wide, and ¼ inch thick. Many colors are available.

21/ **PLASTER CEILINGS**

I always get an odd sort of pleasure out of being able to tell doubting Thomases who profess to be all thumbs that of course they can improve and repair their homes themselves. Man was put on this earth to work with his hands as well as his brain, and it's good to see him being forced by the circumstances of the age into doing just that.

But there are certain jobs the average man should not attempt, and plastering is one of them. It simply takes too much experience.

So all I have to say about plaster ceilings is that if you want them, go out and hire a plastering contractor to put them in for you. The price will be pretty high, but no ceilings are any better.

On the other hand, the maintenance and repair of plaster ceilings require no professional assistance. I defy anyone to protest the work is too difficult. Irksome it may be, but never difficult.

Maintaining plaster ceilings.

In the days when houses were heated with coal, ceilings were filthy. The hot air eddying

from the floor registers carried a load of soot such as you simply wouldn't believe. And naturally the first surfaces encountered by the rising air were the ceilings. In regions where soft coal was favored over hard because it was much cheaper, ceilings became so begrimed so rapidly that fastidious housewives had them cleaned twice a year.

But those were the not-so-good old days. Today almost no one heats with coal, most heating plants are equipped with filters, and the heated air issues from wall registers with enough force to keep it from gravitating straight toward the ceiling. So you hardly ever have to clean ceilings except in the kitchen and, perhaps, bathrooms. This is not to say, of course, that ceilings outside kitchens and bathrooms don't still get dirty, but when they do—perhaps every four or five years—I suspect most homeowners repaint rather than wash them.

In a roundabout way, that's all there is to maintaining plaster ceilings or any other kind of ceiling.

1) Kitchen ceilings should be painted with semigloss or gloss alkyd paint, and washed about once a year or whenever the

film of grease begins to afflict you. Use trisodium phosphate, household ammonia, or any other strong cleansing agent. Rinse immediately so you don't have to climb up and down the stepladder so often.

Repainting is called for only when the ceilings look tired. If you don't change the color, wash and rinse well, roughen gloss paint with sandpaper, and roll on one new coat.

2) Bathroom ceilings should also be finished in semigloss or gloss alkyd paint, and washed when they look dull or mildewed. Use the same cleaning agents as in kitchens, and rinse well. If mildew is present, wash with 2/3 cup trisodium phosphate, 1 quart chlorine bleach, and 3 quarts warm water.

Repaint with alkyd semigloss as necessary.

3) Because ceilings in other rooms are usually painted with flat latex paint, washing is only partially effective in removing accumulated soil. So the sensible practice is to ignore the ceilings until they become so grimy or smudged that you can no longer stand them. Then, after dusting, just roll on a new coat of latex flat paint.

4) Remove cobwebs and dust festoons from all ceilings as you notice them. Use a vacuum cleaner and avoid touching the nozzle to the ceiling. Never try to remove cobwebs or dust festoons with a brush or dust mop, because that leaves indelible black smears on the ceiling.

Repairing plaster ceilings.

While a number of fillers are used to repair plaster, the two you will rely on most heavily are spackle and gypsum board joint compound. In both cases, use the kind that comes ready-mixed in a can. It saves time and mess and produces a smoother, stronger repair.

Small cracks. Scrape the cracks open with any handy sharp tool, such as a beer can opener or screwdriver. It's not necessary to open the cracks very wide or very deep (1/16 inch may be ample) if the plaster is hard, but remove all soft or weak plaster. Then fill the cracks with spackle or gypsum board joint compound; sand smooth when dry, and paint.

I favor spackle for filling individual cracks because one application is usually enough (joint compound shrinks more). On the other hand, I use joint compound for covering a big area that is crisscrossed with cracks because it is softer and spreads better.

Big cracks. Unlike small cracks, most big cracks usually are caused by settlement of the house or warping of the framing members and therefore keep on reopening until the settlement or warping stops. This makes permanent repair an iffy proposition, but your chances of success are improved by a new rubberized paper tape called Super Crack Seal.

Wash the area around the crack, scrape the crack open, and fill with spackle or patching plaster. When the filler has dried, sand it smooth and brush on a thin coat of wallpaper size. Cut the tape to the length of the crack. For a crooked crack, you will need several pieces of tape trimmed at the ends so they butt together. Then immerse the tape in room-temperature water and smooth it over the crack with a damp sponge. When dry, cover with a thin, 4- to 6-inch-wide layer of gypsum board joint compound. Sand smooth and paint.

Holes. Fill small holes with spackle.

Cut out around large holes so the edges are straight up and down or beveled backward. Dampen the edges with water. Then trowel in patching plaster and strike it off flush with the surrounding surface. For a smooth, hard finish, wet a plasterer's trowel or wall scraper frequently with water. For sand-textured plaster, mix the patching plaster with an equal amount of clean fine sand or Perltex.

Very deep holes should be filled in two or

three thin layers; if you try to fill them all at once, the damp patching plaster will sag. To ensure that each new layer sticks to the one below, score the surface of the bottom layer with the point of your trowel or a nail. Apply the new plaster as soon as the previous layer has set but is still damp.

If a hole is more than about 1 foot across, patching plaster is hard to use because it hardens faster than you can work. Use gypsum board joint compound instead if the hole is only about ⅛ inch deep. Level the compound as well as you can with a plasterer's trowel, then draw a steel straightedge or carpenter's square across it to strike it off level with the surrounding surface and trowel smooth once more.

For deeper holes, start with a ready-mixed plaster containing wood fibers or sand. Fill the hole to within ⅛ inch of the surface. When the plaster dries, fill the hole the rest of the way with gypsum board joint compound.

If the lath behind a large hole is broken or rusted out, enlarge the hole until the joists on either side are exposed. Then nail to the joists a piece of plaster lath or steel lath cut to fill the hole, and build up from this with ready-mixed plaster and a final layer of joint compound.

If the lath is missing behind a hole made for a ceiling light or pipe, follow the directions in Chapter 22 for making repairs.

Bulges. If the bulging plaster is loose from the base coats of plaster or plaster lath, knock it out and fill the void as you would an ordinary hole.

Bulging plaster that is sound and firmly fastened to the lath can be handled in the same way. But an easier and equally satisfactory solution is to apply gypsum board joint compound around the bulge until it disappears from view (if not from reality). For example, if the plaster bulge is ½ inch high and 1 foot wide, spread joint compound from the peak of the bulge further and further out to all sides until the bulge is perhaps 4 feet wide but still only ½ inch high.

Powdery plaster. If plaster becomes soft and chalky soon after it's applied, dig it all out and apply fresh plaster. If powdering occurs in old plaster, however, it is usually because there's a water leak somewhere nearby. Stop this first. Then remove the white surface coat of plaster—the one that is powdering. If the brown coats underneath are also crumbling, remove them, and trowel in ready-mixed plaster or patching plaster. If the brown coats are reasonably sound and hard, however, let them dry for several days, then replace the missing white coat with gypsum board joint compound.

22/ GYPSUM BOARD CEILINGS

The great majority of houses built since World War II have gypsum board ceilings (and interior walls). Sometimes I find this cause for regret, since one of the worst problems with these houses—their frightful noisiness—is attributable to the gypsum board in them.

Don't leap to the conclusion that I mean gypsum board is a poor product. On the contrary, it's such a good product that cost-cutting builders can use the thinnest sheets to surface ceilings and walls. But in doing so, they permit sounds to travel through their houses from end to end and top to bottom.

Gypsum board has gained the ascendancy over plaster in ceilings and walls because it is less expensive to install, goes in quickly, and doesn't cause the moisture problems that sometimes plague builders using plaster. Furthermore, the board produces a strong, durable, smooth wall that takes paint and wallpaper beautifully.

Add to these virtues the fact that almost anyone can install it if he has the strength to hoist it into position and hold it while it's being nailed. But that's a big "if" when you're building or rebuilding a ceiling. No matter how strong you are, don't try to undertake this project by yourself. You need at least one strong male assistant or two female assistants.

Types of gypsum board.

Although no one ever seems to mention it, there are several types of gypsum board. About half are used on ceilings.

Standard boards—the kind you always get if you call up a lumberyard and say simply, "Send me six sheets of gypsum board"—are square-edged panels with a thick, dense, fire-resistant core of calcined gypsum sandwiched between two layers of strong paper. The face of the board is made of whitish paper, the back of gray paper. The installed cost is approximately 15 cents a square foot.

The boards come in 4-foot widths and 6- to 16-foot lengths. For resurfacing existing ceilings, ¼-inch-thick panels are used; ⅜-inch for multi-ply construction of ceilings and in

141

single thicknesses for top-story ceilings, ½-inch for most ceilings, and ⅝-inch in ceilings requiring extra soundproofing.

Superior fire-rated boards are similar to standard gypsum boards but more resistant to fire and to the transmission of sound. They are available in ½- and ⅝-inch thicknesses only.

Foil-back boards are covered on the reverse side with aluminum foil to provide thermal insulation and to serve as a vapor barrier. They should be used only in top-floor ceilings, and even here they are of doubtful value now that the cost of heating has risen so sharply, because they don't insulate as well as 4 to 6 inches of fiberglass. (And if you place fiberglass on top of the boards, the aluminum foil is worthless as a heat-stopper.)

Backer boards are low-cost panels used as a base layer in multi-ply ceilings (those covered with two layers of gypsum board to prevent the transmission of sound).

Estimating how much gypsum board you need.

Measure the total length of the ceiling and multiply by the width to find the square foot area. Then divide by the area of the panels you need.

The most commonly used panel size is 4 x 8 feet (32 square feet). I never recommend larger sizes to do-it-yourselfers because they're too heavy and awkward to handle—especially over your head. But if you have a room that's 10 or 12 feet wide, you might prefer to use panels of these lengths because there will be fewer joints to tape.

Other materials you will need are gypsum board joint tape, joint compound, and nails.

Allow 75 feet of tape and 1 gallon of ready-mixed joint compound for every 200 square feet of ceiling surface.

If applying gypsum board to ceiling joists, rafters, or furring strips, use annular-ring nails (also called ring-grooved nails). They have much greater holding power than other nails and are less likely to come loose and pop through the surface of the ceiling. Use 1⅛-inch nails for ¼-inch gypsum board, 1¼-inch nails for ⅜-inch board, and 1⅜-inch nails for ½- and ⅝-inch boards. Allow no less than 1 pound of nails for every 200 square feet of surface.

To resurface an existing ceiling with ¼- or ⅜-inch gypsum board, use 1⅞-inch cement-coated nails at the rate of 1¼ pounds per 200 square feet. For ½- or ⅝-inch gypsum board over an existing surface or new backer board, use 2¼-inch cement-coated nails at the rate of 2 pounds per 200 square feet.

Preparing for an installation.

Gypsum board is applied to ceilings in three ways: (1) directly over the joists or rafters; (2) directly over the existing ceiling surface; or (3) over furring strips.

In a new or unfinished room with joists or rafters exposed, no special preparations are necessary. In an old room, if you plan to apply the gypsum board directly to the framing, you must, of course, tear out the old ceiling and pull all protruding nails. But this approach doesn't make much sense unless the old ceiling is in such hopeless condition that it will be more work to install gypsum board over it than to remove it.

On the other hand, applying gypsum board directly to joists or rafters has one advantage if you're working on a ceiling under the roof: You can staple heavy polyethylene film to the undersides of the framing members as a vapor barrier. No other kind of vapor barrier is as effective.

Applying gypsum board directly to an existing ceiling should be done only if the ceiling is reasonably level. Even so, you should go over the ceiling and knock out areas that bulge downward (don't worry about filling the holes). Level areas that bulge upward by tacking short, thin wood strips across the centers. Lower the boxes for ceiling lights ¼ or ⅜ inch, depending on the thickness of the new gypsum board, so the rims will be flush with the new ceiling surface.

Installing gypsum board over furring strips requires even more preparation and has the added disadvantage of reducing ceiling height about 1 inch. This installation method is called for, however, if the ceiling has an objectionable slope, if pipes are hung slightly below the joists (as in a basement), or if an existing ceiling is so uneven that it would take too much work to level it by the steps described above.

Furring is done with rough 1- x 3-inch boards nailed at right angles to the joists or rafters. The boards, which are generally referred to as strips, are nailed directly to the joists or over an existing ceiling. Use nails that will penetrate the joists at least 1½ inches. One strip must be applied across the ceiling next to each end wall. Space the intervening strips exactly 16 inches from center to center.

The first step prior to installing the furring strips is to check the level of the existing ceiling or joists with several taut cords stretched across the ceiling and a carpenter's level. In a basement with pipes projecting below the joists, stretch the cords across the room just below the pipes. The furring strip installation should then be planned to correct for the slope or unevenness of the ceiling or to compensate for the pipes. Some furring strips may be nailed directly to the ceiling or joists; some may be wedged down from the ceiling or joists; some may be slightly recessed in channels cut through the ceiling surface or notches cut in the bottoms of the joists.

For example, if a ceiling is ¾ inch higher at one end than the other, the easy way to level it is to nail the furring strips directly to the ceiling at the low end and wedge them out from the ceiling with progressively thicker strips of wood as you approach the high end. On the other hand, if the ceiling has a 2-inch slope, it is better to recess the furring strips in the ceiling at the low end so you don't have to wedge those at the high end out so far. The reason for this is that the thicker the wedges used between furring strips and ceiling joists, the less secure the attachment of the furring strips to the joists.

Installing a single layer of gypsum board over joists, rafters, or furring strips.

If an entire room is to be paneled with gypsum board, the ceiling is done before the walls.

The room should be warmed to at least 55° and ventilated sufficiently to get rid of moisture given off by the drying joint compound. However, if the gypsum board is delivered to the house some days before installation starts, it can be stored in a cold room. Lay it on the floor; if stood on edge, it may warp.

Mark the center lines of the joists, rafters, or furring strips on the walls just below the ceilings so you won't have trouble finding them when nailing up the panels.

With a carpenter's square, check whether the corners of the room are square and whether the joists or furring strips are at right angles to the walls. If they are, you should not have to trim the edges of the gypsum board panels to fit the walls. If they're not, some trimming may be necessary—although you won't be able to tell how much until you hold the first panel against the ceiling with the panel edges parallel to the joists, rafters, or furring strips. While the wall-side edges of the panel do not have to lie exactly parallel to the walls, the gap between them should not be greater than the thickness of the panels that will later be used to cover the walls.

Install gypsum board panels with the long edges perpendicular to the joists, rafters, or furring strips. Both ends must be centered on framing members. Put the first panel in a corner and go on from there with full-size panels. Then fill in the gaps in the ceiling with partial panels. On a slanting ceiling, install panels from the bottom up. In all cases, the edges of adjoining panels should be snug but not jammed together. If the panels you use have tapered edges, don't butt a cut edge to a tapered edge, because the joint is hard to conceal.

Have an assistant help you lift each panel into place. Then, while he holds it, slip one or two T-shaped braces underneath. These are made of 2x4s or 2x2s cut 1 inch longer than the finished ceiling height. Nail 1- x 4-inch boards 2 to 3 feet long at right angles to the tops of the 2x4s.

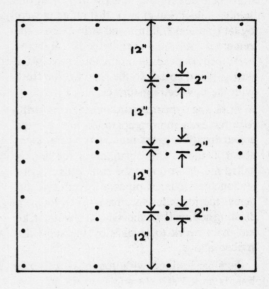

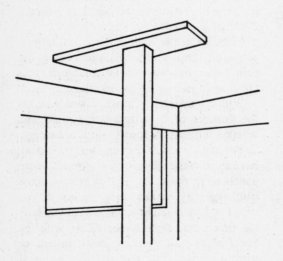

Nail the panels from the centers toward the ends and edges. In standard nailing, the nails are spaced 6 to 8 inches apart along the ends and up and down the intermediate joists. Drive in the nails ⅜ to ½ inch from the panel edges and set each head in a slight dimple about 1/32 inch below the paper surface. Take care not to break the paper around the nailhead and don't use a nail set. While driving each nail, press on the panel alongside to assure that it is drawn up tight to the framing.

In a newer and better nailing system, called double nailing, the nails at the ends of the panels are spaced 7 inches apart, and one nail is used at the long edges where they cross the joists, rafters, or furring strips. But in the middle of the panels, the nails are inserted in pairs. The distance between the center of one pair and the center of the next is 12 inches. The nails in each pair are 2 inches apart. The advantage of this method is that it helps to prevent nail-popping.

Whichever nailing method you use, you will find it helpful after initially affixing each panel to the ceiling to draw lines across it to mark the locations of the joists. Then you won't waste time driving in and pulling out nails that just miss the joists.

Gypsum board panels are cut by scoring the whitish paper on the face with a sharp knife. Use just enough pressure to slice through the paper into the gypsum core. Always use a straightedge to guide the knife for straight cuts, and don't hurry the cutting.

Snap or bend the board backward with your hands to break the core. Then cut through the back paper with your knife and smooth the rough edges with the knife, rasp, or Surform tool.

To cut a piece out of the side or center of a panel, use a crosscut or keyhole saw as well as a knife. (When using a keyhole saw to cut out an inside hole, as for a ceiling fixture, it's unnecessary to bore holes to start the saw. Just jab the sharp point of the blade through the board and begin sawing.)

Installing gypsum board in a double layer.

Multi-ply construction of ceilings improves fire resistance, reduces noise transmission between floors, and helps to produce a smoother, stronger ceiling.

Install the first layers of panels at right angles to the joists, rafters, or furring strips. Use nails and space them by the standard method.

The exposed, or second, layer of panels can be installed in the same direction as the first or across it. In either case, the joints in this layer should offset those in the bottom layer by one joist space or more.

Installation of the exposed panels may be made with nails or adhesive. Nailing is easier. Use cement-coated nails long enough to penetrate well into the joists and apply them by the double-nailing method.

Gluing makes a tighter, more sound-resistant ceiling. The gluing method depends on the adhesive used, so follow the directions on the container. Generally, the adhesive is applied to the backs of the surface panels with a notched trowel or caulking gun. The panels are then positioned on the ceiling and firmly bonded by rapping them all over with a rubber mallet or with a board and hammer.

To hold the panels while the adhesive sets, drive common nails into the framing at 16- to 24-inch intervals. The nails should be long enough to penetrate the framing at least ¾ inch. Place scraps of gypsum board under the heads to serve as washers that will protect the ceiling surface. When pulling the nails after the adhesive dries, slip a thin block of wood under the hammerhead. Dimple the nail holes with your hammer before filling them with joint compound.

Installing gypsum board over an existing ceiling.

Just because there is a flat surface that serves as a base for the new gypsum board doesn't mean that the edges of the gypsum board panels can fall just anywhere. On the contrary, if the panels are installed at right angles to the joists, the ends must be centered over and nailed to the joists. Similarly, if the panels are installed parallel to the joists (as is permissible), the long edges must be centered over and nailed to the joists.

Actual installation is made in the way already described. Mark the locations of the joists on the side walls and/or existing ceiling so you won't waste time hunting for them when nailing up the new panels. Be sure to use nails long enough to go into the joists and handle them either by the standard or double-nailing method.

Hanging gypsum board panels from resilient channels.

Resilient channels are preformed strips of thin steel which allow you to attach gypsum board to joists in such a way that there is no direct contact between board and joists. As a result, sound vibrations transmitted by the joists are dissipated before reaching the gypsum board, and the noise of people moving around and talking upstairs is hardly audible in the room below.

The channels are installed at right angles to the joists in parallel rows 24 inches apart. Fasten them to the joists with screws driven through prepunched holes.

The gypsum panels are applied perpendicular to the channels with the end joints of adjacent panels staggered. Attach the panels to the channels with special screws supplied by the panel manufacturer. Space the screws 12 inches apart along the ends and through the field of the panels. Around ceiling fixtures, 2- x 4-inch blocks must be nailed between the joists to support the panels, and the outlet boxes should be lowered so the rims are flush with the ceiling surface.

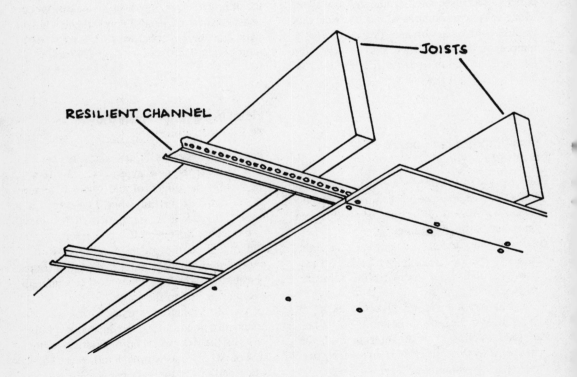

Finishing joints.

All joints between panels are finished in the same way regardless of how the ceiling is constructed. The job will take perhaps an hour a day for three consecutive days. Don't try to rush it: The more carefully you work, the smoother the ceiling will be. To apply the joint compound, use nothing smaller than a 4-inch wall scraper or a broad knife. Wider scrapers (first a 6- or 8-inch, then an 8- or 10-inch) simplify the application of the second and third coats of joint compound. You can use the smooth (unserrated) edges of a rectangular flooring trowel instead of wide scrapers.

Finish the joints in any sequence you like, but once you start on a joint, complete it before shifting to another joint. If you don't do this, the joint compound, which dries quite rapidly, will start to harden before you've smoothed it.

The first step is to spread an even ribbon of gypsum board joint compound from one end of the joint to the other. Center it over the joint and don't make it any more than 1/16 inch thick.

Starting at one end of the joint, center the paper reinforcing tape over it and press it into the compound as you gradually unroll the roll. Be sure the tape is firmly embedded at all places, but don't bear down so hard that you squeeze the compound out from under it. If it starts to wrinkle, pull it off and press it down again. If it veers away from the joint, tear it in two and butt a new length of tape to the end. When you reach the end of the joint, hold the edge of the scraper against the tape and tear off the roll.

Having embedded the tape, return to your starting point and cover it with a skim coat of joint compound. Smooth this out as well as possible to remove ridges and uneven places.

Complete the other joints in the same way.

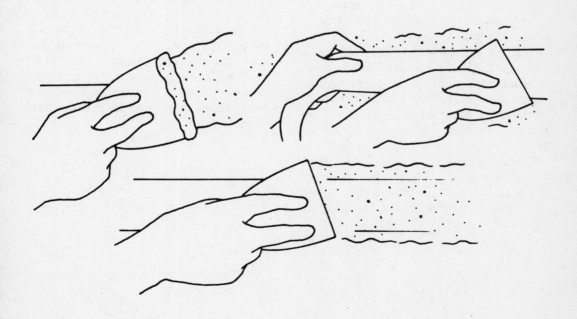

Right-angle joints between ceiling and walls do not have to be taped if you plan to cover them with a molding. If not, spread compound into them and about 2 to 2½ inches out on the ceiling and walls. Cut the tape to the length of each wall and fold it in half lengthwise. Then embed it in the compound and cover with additional compound.

Joints of less than 90° between a flat ceiling area and slanting ceiling area or between a slanting ceiling area and a wall can be taped in similar fashion. However, a special tape made with a thin ribbon of reinforcing metal is a little easier to handle.

When all the joints are filled, apply joint compound over exposed nailheads, dents, and holes.

Let the compound dry thoroughly (it turns from a medium wet-looking brown to a dry-looking yellow-brown). This will take twelve hours or more. Then sand the compound with medium sandpaper to remove any roughness. A reciprocating electric sander saves time and energy. Then apply a second thin coat of joint compound in ribbons about 8 inches wide—2 inches to either side of the first ribbons. Feather the edges.

Let the compound dry, and sand again. Nailheads and holes need no further covering. But a third thin coat of compound must be applied over the joints. Make the ribbons 10 to 12 inches wide, and take extra pains to get them as smooth as possible and to feather the edges. When dry, sand once more. Examine the joints for depressions or scratches by looking at them and by rubbing your hand across them, and either sand them smooth or apply a little more compound. Finally, dust the entire ceiling thoroughly.

Covering joints with battens.

Although most homeowners prefer smooth, unbroken ceilings made with joint compound and tape, joints between panels are sometimes covered with wood battens. Such treatment adds an attractive, geometrical texture to the ceiling—but only if the ceiling can be divided into sections of equal size. For example, a 16- x 20-foot ceiling lends itself perfectly to batten joints because each framed section measures exactly 8 x 4 feet. On a 16- x 19-foot ceiling, on the other hand, you would have eight 8- x 4-foot panels and two 8- x 3-foot panels.

The battens used should also be attractive. Ordinary flat 1- x 3- or 1- x 4-inch boards don't altogether meet this requirement. A much better effect is gained if you nail small ornamental moldings along the edges of the boards. This obviously increases the amount of work you must do.

To make a batten installation, there is no need to tape the joints between gypsum board panels. Just center the battens over the joints and nail them to the joists with finishing nails.

First, nail battens around the four sides of the ceiling. Then cover the joints running across the narrow dimension of the ceiling. Use boards that are long enough to extend from wall to wall. Then cover the lengthwise joints. Finally, tack the moldings to the edges of the boards.

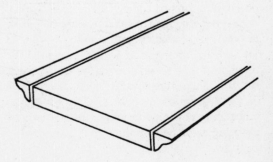

Finishing the ceiling.

Gypsum board ceilings are usually painted with flat latex paint or—in kitchens and bathrooms—with semigloss or gloss alkyd. In either case, use a latex primer for the first coat.

If the ceiling is to be covered with wallpaper or other flexible wall covering such as vinyl, you should also give it a prime coat of paint so you can later remove the paper, or whatever, without pulling off the paper surface of the gypsum board. Latex primer is generally recommended for this, but after a recent unfortunate experience in stripping off wallpaper, I now strongly favor an alkyd primer.

Repairing gypsum board ceilings.

If annular-ring or cement-coated nails were used to install a gypsum board ceiling, nail-popping—the scourge of most gypsum board installations—would never occur. But speculative builders have a bad habit of using other kinds of nails, so you may find that ceilings they put up in your house develop what looks like a geometrical case of hives. Happily, the disease is curable. Hammer in the loose (popped) nails, and drive in an additional nail about 2 inches away from each one. Butter with joint compound, sand, and repaint.

Cracks that occur at joints between gypsum boards are simply scraped open and filled with joint compound or spackle. If they recur, cut out the joint compound on either side of the tape and scrape it off from over the tape. Should the tape be wrinkled or loose, cut that out, too. Then drive new nails along the edges of the gypsum panels and refill the joint by embedding new tape in joint compound and covering with two or three layers of compound.

Holes are rarely a problem in ceilings, but you never can tell when little Willie will shoot an arrow through a ceiling or when a roof leak will damage a ceiling to such an extent that you must make a hole in it to fix it. Or you might even decide to take out a ceiling fixture. So be prepared.

Fill small holes with joint compound or spackle. If the filler doesn't hold, wad steel wool into the hole and apply the filler over it.

Holes between 1 inch in diameter and the size of an electric outlet are filled by wetting the edges of the gypsum with water. Cut a piece of cardboard slightly larger than the hole, poke a small hole in the center and thread a piece of string through it. Knot this on the back side. Then push the cardboard diagonally through the ceiling hole and pull it down against the back of the ceiling with the string. Fill the hole to half its depth with patching plaster while pulling on the string. When the plaster sets, cut the string off flush with the plaster and fill the hole the rest of the way.

Large holes are filled with patches cut out of a waste scrap of gypsum board. First trim around the hole with a saw to make a square or rectangular opening. Add 2 inches to both dimensions of the opening and cut the gypsum board scrap to this size. This is the patch.

Turn the patch back side up, measure in 1 inch from all edges, and draw an outline of the opening in the wall. Cut across the patch along these lines with a knife and bend the edges backward to break the gypsum core. Then trim the core edges away from the surface paper. You now have a plug with flanges on the front.

Trowel a thin layer of joint compound around the sides of the opening. Push the plug into the opening and smooth the flanges into the compound. Then apply a little more compound over the edges of the flanges. Until the compound dries, hold the plug in place with a T-support like that used in erecting a ceiling.

Then apply a thin layer of joint compound over the entire patch and well out on the surrounding surface. Feather the edges. Finish the job the next day with a still wider coat of compound.

If a large section of ceiling begins to sag as a result of a leak in the roof or pipes above, cut out the entire area in a rectangle. Then enlarge the hole as necessary back to the adjacent joists. Nail 1- x 3-inch boards to the sides of the joists flush with their bottom edges. Cut 2x4s to fit snugly between each pair of joists, and nail these across the ends of the hole. The blocks must be positioned so they are centered under the edges of the cut gypsum board panels. Nail the panels to them.

Out of gypsum board of the same thickness as the ceiling cut a rectangle to fill the hole, then nail it around the edges to the 2x4s and 1x3s. Nail it also to any intermediate joists. Then tape the joints.

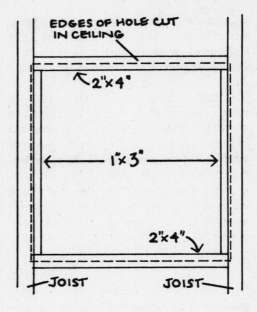

23/ PLANK CEILINGS

One of the features of many of today's open-plan houses is a plank ceiling. These are especially common in houses in which the ceilings follow the roof line. Generally the planks are laid atop the rafters or ceiling joists to produce beamed ceilings similar to those in early colonial homes. But sometimes the planks—which in this case are really boards—are nailed to the bottoms of the rafters or joists. Whichever method of installation is followed, the effect is usually delightful.

Wood always adds a special sense of warmth to a room. This is particularly true if it is given a natural finish so you can actually see the pores, grain, and flaws in the wood. But even when painted, the feeling is similar. That's reason enough for building plank ceilings. But additional arguments are not hard to find: A plank ceiling provides good thermal insulation. It requires little care. And if a leak should develop, it doesn't require extensive ceiling repairs.

Types of plank.

Ceiling boards are 1-inch boards nailed to the undersides of ceiling joists and rafters.

In the lumber industry, the term *ceiling board* refers to boards that are machined to various simple patterns for use specifically on ceilings, but any kind of tongue-and-groove board can be used in the same way. The boards can have square edges, bevel edges, beaded edges, and so forth; the surfaces can be smooth, grooved, beaded, and so forth. Generally, the boards are cut from fir, redwood, and other softwood, but if you prefer something else, all you have to do is ask for it. Usual board widths are 6 and 8 inches, but again, there is no reason why you shouldn't use anything that suits the room and house.

Deck planks are thick (up to 5-inch) tongue-and-groove planks used in plank-and-beam construction. Installed on top of rafters and ceiling joists, they serve not only as the ceiling surface but also as the base on which insulation and roofing are laid. They come in 6- and 8-inch widths. Made of spruce, fir, pine, cedar, larch, or redwood, they most often have smooth faces and beveled edges, but other patterns are available.

Composition planks are made of compressed wood fibers in 2- to 4-foot widths and 1½-, 2-, and 3-inch thicknesses. Like solid wood deck planks, they are installed above rafters and ceiling joists; they serve as ceiling surface, roof deck, insulation, and vapor barrier. Obviously they don't look like wood, but seen from several feet below, they have an attractive textured surface that holds paint well.

Estimating your needs.

Determine the square footage of the ceiling or roof deck. Then, if you're going to use solid wood planks or boards, let the lumberyard figure how many board feet you will require. Note that unless you specify the lengths of the pieces you want, you will receive a random assortment.

To order composition planks, divide the roof area by the area of the selected planks. Specify the length desired (this ranges from about 4 to 12 feet).

Paneling a ceiling with boards.

Board paneling must either be nailed directly to the joists and rafters or to furring strips. The former installation requires no advance preparations, but you must install the boards either at a 90° or 45° angle to the framing members.

Use of furring strips allows you to install the boards over an existing ceiling but requires that you lay the ceiling boards parallel to the joists or rafters. Use 1- x 3-inch strips and space them 16 inches on centers. For how to install them, see Chapter 22.

Have the ceiling boards delivered several days before installing them and pile them indoors on the floor. If you want a ceiling with a natural finish, stain the boards prior to installation, and make sure you coat the edges, so that if the boards contract, the joints won't show up as white lines. If the ceiling is to be painted, prime the knots with a stain-killer and apply an alkyd primer to the face and edges of the boards. Then sand all boards, whether painted or stained.

On a flat or gently sloping ceiling, start installation at one side and work across to the other. Measure the width or length of the ceiling and divide by the actual face width of the boards to make sure that the last board installed will not be of sliver width. If it will be, add the width of the sliver to that of a full-size board, divide in two and cut both the first and last boards to the resulting dimension.

On a steeply sloping ceiling, install boards from the bottom up, so that if you end up with a narrow board, it won't be obvious to the eye.

In a new room, the ceiling is built before the walls. If you fit the boards to within ¼ inch of the wall framing, the wall surfaces will conceal the gaps and there will be no need to cover the joints with moldings. On the other hand, if you're paneling the ceiling in an existing room, it's almost impossible to eliminate all gaps between the ceiling and walls except by scribing* each board carefully to the walls. This is a tedious chore. The

---

* Scribing means to make a line parallel with another line and then to cut along it. In this case, the "other line" is the wall. To scribe a ceiling board to the wall, nail it temporarily to the rafters parallel to the wall and about ½ inch away from it. Open a pair of dividers or a compass slightly wider than the widest point in the gap. Hold one leg against the wall and the other leg on the board and draw them lengthwise along the board. This will mark on the board a line conforming to the contours of the wall. Trim the edge of the board to this line with a saw, drawknife, or plane so it will butt tight against the wall.

easier solution is simply to fit the boards as well as you can and install moldings around the edges of the ceiling.

Apply the boards with 2-inch finishing nails. If a ceiling is to be finished with paint or an opaque stain, the nails can be driven through the face of the boards, countersunk, and covered with spackle. On a ceiling with a transparent finish, however, drive the nails diagonally through the tongue of each board. Only the first and last boards are face-nailed. In all cases, to prevent movement, nails should be driven through all boards into each of the joists or furring strips they cross.

As in a floor made of wood strips or planks, the tongue of each new board must be set snugly in the groove of the adjacent board. Hold a block of wood against the groove edge of the board and hammer the board into place. If you don't get a tight fit because of a warp, cut out the crooked section and use the straight piece only.

All end joints must be made over framing members or furring strips.

Installing wood or composition deck planks.

Strictly speaking, when you use wood or composition planks, you are building a roof. The ceiling, which is the underside of the roof, is incidental. And, of course, the decking itself is only one part of the roof.

In a few cases, when purlins are installed between roof beams, the planks are laid up and down the roof. But generally planks are laid across the beams, rafters, or joists and nailed directly to them. Follow the nailing schedule recommended by the decking manufacturer. This ordinarily calls for nailing each plank to each framing member and also toenailing the planks to one another.

Work from the bottom of the roof up to the peak, and don't worry how the planks look

from the underside. If only a sliver of the first plank shows at the base of the ceiling, there's little you can do about it.

End joints of composition planks are centered over the framing members. Solid planks, which are tongued and grooved at the ends as well as along the edges, are joined either over the framing members or between them. In the latter case, the planks must rest on at least one support, and the joints in adjacent rows of planks must be well staggered.

When the roof deck is completed, it is covered with rigid insulating boards and roofing.

Finishing plank ceilings.

Pressed wood planks are usually painted with a coat of alkyd primer followed by a coat of alkyd flat, semigloss, or gloss interior enamel.

The same painting system is used on solid wood planks and ceiling boards. But as a rule, these ceilings are given a transparent finish or are not finished at all. (see Chapter 5.)

If you want to change the color of the wood under a transparent finish, stain the ceiling boards before they are installed. You can also follow this procedure with deck planks, but before doing so it's a good idea to investigate what the decking makers offer in prefinished material. One of them supplies decking in about sixteen colors, including browns, reds, greens, and charcoal.

Maintaining and repairing a plank ceiling.

Ceiling that are painted or varnished need only the simplest care. Vacuum off cobwebs and dust festoons. Wash when the ceilings become grimy or greasy.

Ceilings that are unfinished or given only a

light protective coating of wallpaper lacquer are a bit more troublesome—but only if you make the mistake of doing this in kitchens or bathrooms. When vacuuming, take great care not to smear the dirt on the wood. To remove grime, sand the wood with medium-fine sandpaper or coarse steel wool. Work with the grain. Grease must be attacked with paint thinner, and all you can do is trust to luck that you get it off without rubbing it into the pores. Failing this, hard sanding is in store for you.

Repairs are minimal. In fact, no matter how I cudgel my memory, I can't think of any problems I have ever had to cope with in a wood ceiling. Of course, if you have a leak and don't attend to it promptly, you may end up with a case of rot or a mess of hard-to-obliterate stains, but why borrow trouble?

24/ CEILING TILES

There are two points that must be made clear at the outset:

First, although most people seem to think that all ceiling tiles are acoustical tiles—certainly they refer to them as acoustical tiles—the fact is that a great many are not. They're just ordinary composition tiles without any sound-deadening characteristics whatever.

Second, acoustical tiles—the real honest-to-goodness acoustical tiles—do not stop the transmission of sounds from room to room and floor to floor. They are designed only to absorb and muffle noises made within rooms. For example, when you have a birthday party for eight-year-olds in your family room, the acoustical tiles on the ceiling soak up the noise the kids make so the grown-ups supervising the party don't go totally out of their minds. But unfortunately for anybody who is sick in another part of the house, the tiles have no dampening effect on the racket emanating from the family room.

So what's so wonderful about ceiling tiles?

Nothing really. They just happen to have been given a bigger advertising and publicity buildup than many other ceiling coverings.

Ceiling tiles are simply big, flat squares or rectangles made of cellulose or mineral fibers. (There are also tiles made of perforated metal, but these are rarely if ever used in houses.) The standard tile is a 12-inch square roughly ½ inch thick. There are also several larger squares and rectangles (to a maximum of 2 x 4 feet), which are generally called panels rather than tiles. There are even imitation wood planks 4 feet long and approximately 5, 6½, and 8 inches wide.

True acoustical tiles have a porous texture or are actually perforated with hundreds of small holes. Nonacoustical tiles are of much more solid texture. Both are available in a wide range of two-dimensional patterns and three-dimensional sculptured designs. Most come with a factory-applied finish.

Ceiling tiles have become popular not so much because some of them control sound but because they form a reasonably attractive surface that is very easy to install at a low price. (Inexpensive tiles sell for a little less than 20 cents a square foot.)

They are also useful for lowering high ceilings because they can be installed in light-weight metal frameworks that are suspended at any height above the floor.

In suspended ceilings, tiles also provide quick access to plumbing, heating, and electrical components, and they permit you to change your lighting system without major rewiring.

Selecting ceiling tiles.

There are several points to consider when selecting ceiling tiles:

Cost. I don't have to go into this. Just don't let it dominate your thinking.

Appearance of tiles. Your likes and dislikes are, of course, personal. The one thing that especially worries me about the great majority of ceiling tiles is that they are so heavily textured that they are difficult to clean. Admittedly, this is not a serious drawback if you are using tiles in a living or sleeping area. But if you put them in the kitchen or bathroom, choose the smoothest-surfaced design you can find.

Whether you need acoustical or non-acoustical tiles. If it's the former, be sure to check the Noise Reduction Coefficient of the available tiles before making your choice. (see Chapter 3.)

The method of installation. If the tiles are to be used to resurface an existing ceiling, buy 12-inch squares because they are the easiest to glue to an uneven surface. Over furring strips, use 12-inch squares or "planks." In suspended installations, large panels are preferable.

Estimating your needs.

If you use standard foot-square tiles, measure the length and width of the ceiling in feet and multiply the figures. Order about 2 percent more tiles than the answer calls for.

To order other sizes of tile, multiply the length of the room by the width in feet to get the square-foot area, then divide by the square-foot area of the tiles. Order several extras to allow for waste.

Ordering components of the metal framework in which tiles are installed is more difficult because the framing systems vary from manufacturer to manufacturer. Sears, Roebuck and Company, for instance, gives very clear directions for figuring out what you need. But if you're buying some other brand of tile, you should draw an accurate plan of the room, take it to the tile dealer, and throw yourself on his mercy.

Resurfacing an existing ceiling.

As noted, resurfacing should be done with 12- x 12-inch tiles. The ceiling must be sound and level. Ignore small holes, but fill those that are more than half the size of the tiles. Fill hollows by troweling in gypsum board joint compound and smoothing roughly (but you need not sand).

Lay out the installation as you would a resilient tile floor (see Chapter 10). Locate the center of the ceiling and strike chalk lines across it. Find the distance in feet to the ends and sides of the room, and readjust the chalk lines so the border tiles at the ends of the room are approximately the same width as those at the sides.

Starting at the center of the ceiling, install the tiles over one-quarter of it at a time. Daub the adhesive specified by the manufacturer on the back of each tile. The daubs should be roughly 1½ inches across and ⅛ inch thick.

Put one daub at each corner and a fifth in the middle.

Press the first tile on the ceiling and hold it by driving a couple of ⅜- or 9/16-inch staples through the flanges. Succeeding tiles are installed by interlocking the edges with the adjacent tiles and pressing down firmly. The tiles should be snug but not forced together. Driving staples through the flanges of occasional tiles will help to hold them until the adhesive sets.

The border tiles should fit snugly against the walls if you don't want to cover the joints with moldings. To fit and cut the tiles, follow the directions for fitting and cutting parquet flooring (see Chapter 8). Use a fine-toothed saw or sharp knife and hold the tiles face up.

To fit a tile around an electrical box, rub blue chalk on the box rim and press the tile against it. Then poke a hole through the outlined area and cut the opening with a jigsaw or keyhole saw.

To fit a tile around a vertical pipe, measure the distance from the center line of the pipe to the edge of the previously installed tile, and transfer this measurement to the tile. Cut the tile in two, and notch a half-circle in each piece for the pipe.

Tiles are usually laid so the crosswise joints, like the lengthwise joints, are in straight rows. You can, however, use a bricklike running bond in which the cross joints are centered on the tiles in the adjacent rows.

Applying tiles to furring strips on an uneven ceiling or over an unfinished ceiling.

Use 12-inch squares or narrow 4-foot planks for this installation.

Check the level of the ceiling (see Chapter 4). The furring strips, made of either 1- x 2- or 1- x 3-inch boards, should be installed so the new ceiling will be level. Notch them into low points in the ceiling; wedge them down from high points.

If you're putting up square tiles, find the center of the ceiling. The center point should then be adjusted so the border tiles are of approximately equal width.

The furring strips are nailed to the joists at right angles. Position them so they will be centered under the edges of the tiles—that is, 12 inches apart from center to center.

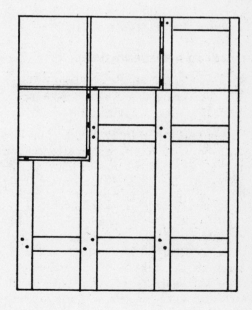

Staple the tiles to the furring strips with 9/16-inch staples driven through the two flanges on the edges of the tiles. In the flange that is centered on a furring strip, space the staples about 4 inches apart. In the other flange, drive one staple into the adjacent furring strips. Interlock the edges of the tiles.

At the borders and in corners, cut the tiles to fit tightly in position and fasten to the furring strips with staples. You may also need to drive finishing nails through the faces of the tiles. Countersink the nailheads.

If you use tile "planks," it is not necessary to find the center of the ceiling. Install the first row of planks next to one of the walls and work across the ceiling to the opposite wall. The furring strips are nailed at right angles to the joists and spaced 12 inches on centers. The planks are then stapled across the furring strips, with one staple driven through the flanges into each strip. End joints between planks must be centered on the furring strips and staggered in adjacent rows.

Installing a suspended ceiling.

Suspended ceilings made with large (usually 2- x 4-foot) panels are hung at least 3 inches below the joists or existing ceiling and may, of course, be dropped much lower. No ceiling preparation is required.

The suspension systems vary slightly; consequently, you must follow the manufacturer's instructions. The general installation procedure, however, is as follows:

First, determine the height of the ceiling and snap a level chalk line around the walls of the room. Then nail the metal wall moldings, which are part of the suspension framework, to the walls at this height. Drive one nail into each stud. On a concrete wall, use masonry nails or adhesive-backed screw anchors and space them 24 inches apart.

The metal framing pieces installed across

the ceiling consist of 10- or 12-foot main beams and 4-foot or 2-foot cross-Ts. The beams are installed perpendicular to the ceiling joists and usually support the ends of the panels; the cross-Ts support the panel edges. This arrangement means that, if you use 2- x 4-foot panels, they are parallel to the joists, and this in turn means that in most rooms they are at right angles to the long sides of the ceiling. If you want the panels to run lengthwise of the ceiling, you must use 2-foot cross-Ts rather than the 4-footers normally provided, and the edges of the panels must rest on the main beams.

Find the center of the ceiling and snap a chalk line across it lengthwise on the joists or ceiling. Then snap parallel chalk lines to either side. The spacing of the lines depends on how you want to install the panels. It should be 4 feet if the panels are to parallel the joists, 2 feet if they are to be perpendicular to the joists.

Fasten beam hanger wires to the sides of the studs over the chalk line. Hang the wires from

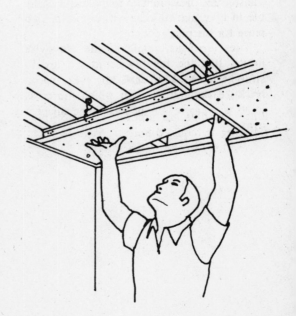

large staples or screw eyes. The wires should be 6 to 8 inches longer than the space between the joists and suspended ceiling. Space the wires about 4 feet apart through the center of the ceiling. No wires are needed at the walls since the beams rest on the wall moldings.

Cut the main beams so the cross-Ts will intersect them at the edges or ends of the panels. If the ceiling is longer than the beams, splice two beams together with one of the splicers provided in the suspension system or, lacking these, with nuts and bolts. Then set the ends of the beams on the wall moldings and hang them from the wires. Check each beam in several places with a carpenter's level.

When all the beams are in place, install the cross-Ts, working outward from the central beam. Then tip the full-size panels up through the metal framework and drop them into place. Cut the border panels to fit, and set them in place.

If you set lights into the ceiling, use those specified by the ceiling manufacturer. Some are mounted on the old ceiling or between the joists, others rest in the suspended ceiling framework, and still others are mounted just below the suspended ceiling.

Finishing and maintaining ceiling tiles.

Since the tiles are prefinished, no further finishing is required unless you want some color other than the white supplied.

Vacuum tiles like any other ceiling material to remove cobwebs and dust festoons. Wash when soiled with a mild detergent solution or water.

The tiles can be painted and repainted as often as you wish, even though they are designed for sound-deadening. Use either latex or acrylic paint, and brush or roll it out well on acoustical tiles. (Since nonacoustical tiles are not porous, it doesn't make any difference how thick a coating you apply.)

Repairing a tile ceiling.

When tiles are damaged in any way, replace them. To remove them from the ceiling, make a cut through the center and work from this toward the edges as you dig out the pieces. To set in new tiles, you will have to cut off some of the flanges. Then glue the tiles to the old ceiling with the proper adhesive or with bathtub caulking, or nail them up with finishing nails and countersink the heads.

If tiles are loose or bulging, nail them; the chances are that you can't get at the underside or flanges to use adhesive or staples.

Panels hung in a suspended ceiling should not warp. But if they do, remove them from the framework, set them concave side up on a floor, and weight them down with bricks. Sprinkling lightly with water should help if the tiles are made of cellulose materials.

25/ **LUMINOUS CEILINGS**

I have a rather wistful note from a representative of one of the companies that used to push luminous ceilings. "Luminous ceilings are not being recommended as in the past. It's not an economical way to light a room."

OK, I won't urge luminous ceilings on you either. But I can't ignore them, because many people are going to put them in whether there's an energy crunch or not. This won't help to solve our desperate oil shortage. But the enthusiasm for luminous ceilings is understandable.

In the first place, no other kind of light fixture (a luminous ceiling is, after all, nothing but an overgrown light fixture) delivers such a high level of glare-free illumination so economically. This is what makes luminous ceilings so desirable in kitchens, bathrooms, and laundries where you need almost all the light you can get.

In the second place, whether a luminous ceiling spans a room from wall to wall or occupies only a smallish section in the center, it's an extremely decorative unit for family rooms, living rooms, dining rooms, halls—not to mention kitchens and bathrooms.

160

What exactly is a luminous ceiling?

It's a fixture made of several parallel rows of fluorescent tubes mounted on a white reflecting surface and directing light downward through a diffusing panel of translucent plastic or openwork. It can be suspended below a new or existing ceiling surface or constructed between the joists in a new or existing ceiling. The cost of the simplest homemade unit comes to roughly $5 a square foot.

Basic rules for constructing a luminous ceiling.

These sound more complicated then they are:
1) Since the minimum height of a ceiling constructed under Federal Housing Administration standards is 7½ feet, a luminous ceiling must also be at least 7½ feet. An 8-foot height is much better.
2) The ideal luminous ceiling is one forming a large panel of light that is interrupted only by the framework supporting the diffusers. Such a panel can be created only by eliminating barriers, such as joists, be-

tween the rows of fluorescent tubes. But fixtures containing barriers need not be ruled out as long as you recognize that the barriers will create shadowy black lines dividing the lighted surface into strips.

3) To provide a uniformly lighted surface without bright streaks, the distance from the center line of the fluorescent tubes to the upper surface of the diffusing panels should be 10 to 12 inches.

4) The spacing between rows of tubes also contributes to the uniformity of the lighted surface. It should not exceed 1½ times the distance between tubes and diffusing panels. This means that if the tubes are 10 to 12 inches above the diffusers, the spacing between the centers of the tubes should be 15 to 18 inches.

5) All surfaces within the light fixture must be painted flat white.

Obviously, before installing a luminous ceiling you must analyze these rules to determine (1) whether an installation is feasible and (2) if feasible, how the ceiling should be built.

First, measure the height of the existing ceiling. (This assumes that you're remodeling a room. If you're adding a new room, the ceiling height is determined—at least in part—by the dimensions of the luminous ceiling.)

If the height exceeds 8 feet 7 inches, you can install a large uninterrupted light panel under it. (Minimum FHA ceiling height—7 feet 6 inches—plus minimum distance from tubes to diffusing panel—10 inches—plus the depth of a fluorescent fixture—3 inches—equals 8 feet 7 inches.)

On the other hand, if the ceiling height is less than 8 feet 7 inches, you may or may not be able to install an unbroken panel of light. But you definitely can have a luminous ceiling divided into strips.

If the ceiling height *plus the depth of the joist space* is less than 8 feet 5 inches, construction of a luminous ceiling is impossible.

Installing a luminous ceiling in a room more than 8 feet 7 inches high.

This is the simplest kind of luminous ceiling to install since you don't have to change the existing ceiling except to make holes for the wiring. The lights required—inexpensive, white-finished, 48-inch fluorescent channels each holding a single 40-watt tube—are available from any electrical supplies store. The metal framework and diffusing panels that comprise the ceiling below the lights are available from firms calling themselves "illuminated ceiling systems specialists."

The size and shape of the luminous ceiling you put in depends on the level of illumination called for, the design of the room, and the size of the available diffusing panels (normally 2 x 2 or 2 x 4 feet). The ceiling may extend from wall to wall or it may be a comparatively small light panel surrounded by a conventional opaque ceiling surface. You need the same kind of metal supporting grid in either case. This consists of wall moldings that support the perimeter of the ceiling, and main beams and cross-Ts forming the central framework.

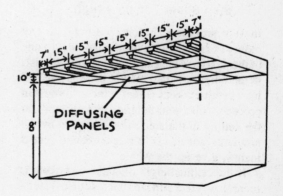

The first step in building a luminous ceiling is to mount the fluorescent channels on the existing ceiling. If the surface is noncombustible, fasten the channels directly to it with screws driven into the joists or with toggle bolts if the channels are between joists. On a wooden surface, pieces of asbestos-cement board must be inserted between the ceiling and the channels to prevent possible combustion. Install the channels end to end in straight parallel rows. The rows should extend to within at least 7 inches of the edges of the diffusing panels underneath. The outside rows of fixtures should also be within at least 7 inches of the edges of the diffusing panels, and the rows between should be spaced 15 inches from center to center.

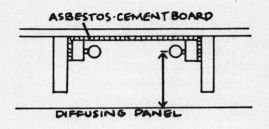

Paint the ceiling and all surfaces surrounding the fluorescent channels flat white.

Build the new light-transmitting ceiling 10 inches below the center line of the fluorescent tubes. For precise installation directions, read the instruction sheet issued by the ceiling systems specialist. For a generalized description of the procedure, see Chapter 24.

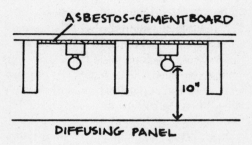

Installing a luminous ceiling in a room less than 8 feet 7 inches high.

In this situation, before putting out any money, bore a hole in the ceiling next to one of the joists and measure the depth of the joist. If the height of the existing ceiling plus the joist depth exceeds 8 feet 5 inches, you can proceed with the installation by knocking out the ceiling in the area to be occupied by the luminous panels. If it's less than 8 feet 5 inches, give up the whole idea.

If the ceiling height plus the joist depth is more than 8 feet 5 inches but less than 8 feet 7 inches, you must install two rows of 4-foot fluorescent channels between each pair of joists in order to create a high enough level of illumination in the joist spaces to kill the

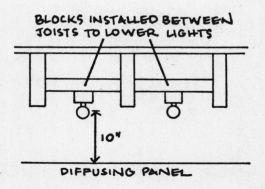

bright streaks the tubes would otherwise make on the ceiling surface. Screw the channels to the sides of the joists directly under the subfloor. Place strips of asbestos-cement board between the channels and joists and subfloor.

If the ceiling height plus joist depth exceeds 8 feet 7 inches, you need only one row of fluorescent channels between each pair of joists. The easiest way to install the channels is to screw them to the underside of the subfloor after it has been covered with asbestos-cement board. However, the joists will then create shadow lines that divide the diffuser visually into strips; consequently, this sort of installation should be made only when the ceiling height plus joist depth is just a little more than the 8-foot-7-inch minimum.

If possible, the fluorescent channels should be attached to blocks of wood nailed between each adjacent pair of joists. The lower the blocks are placed, the less the joists interfere with the spread of light and the more uniform the light on the surface of the diffuser.

A second advantage of lowering the fluorescent channels in the joist spaces is that the bridging can usually be more easily accomodated. It should not, of course, be removed. But if the channels are mounted high in the joist spaces, the bridging must either be installed under them or between them. In either case, it creates dark spots on the diffuser.

Once the fluorescent channels are placed in the joist spaces and the spaces are painted white, the diffusing panels can be installed in one of two ways. The simpler but more costly procedure is to suspend a prefabricated metal ceiling grid from the joists. The alternative—if the ceiling is to be hung just a few inches below the joists—is to fur down the joists with boards nailed along the bottom edges and to rest the diffusing panels (which are cut to the width of the joist spaces) on moldings or metal strips secured to the bottom of the furring

Selecting diffusing panels or grilles.

The selection of diffusers for a luminous ceiling must be made with care because they can spell the difference between a good and bad installation. If they are too opaque, the light output is reduced. If they're too transparent, the bright lines of the fluorescent tubes show through clearly. Then, too, you must think about the color-fastness, weight, and strength of the diffusers, how easily they can be cleaned, and their decorative effect.

Unfortunately, there are no rules the layman can follow in making his choice. The only reliable way to determine whether a diffuser controls brightness well is to test it under a bank of fluorescent ceiling fixtures. Beyond that, you must simply accept the fact that, next to glass, acrylics change color less than other diffusing materials. And, of course, all plastics are lighter, more maneuverable, and less breakable than glass. On the other hand, glass is a little easier to clean than plastic.

The choice between solid and grilled diffusers depends largely on which strikes you as more attractive. Although the former are available with many interesting designs (often colored) molded into the surface or sandwiched between layers of plastic, the latter have a bolder, more interesting texture. The latter also ventilate the space above, collect less dust, and, in a few cases, help deaden sounds within the room. On the other hand, if you use a very open grille, such as an egg-crate design, illuminating engineers generally recommend that a solid sheet of translucent plastic be placed above it so you cannot see the fluorescent tubes when you look up.

Maintaining and repairing a luminous ceiling.

The only maintenance a luminous ceiling requires is periodic replacement of the fluorescent tubes and thorough cleaning about every two months.

Tube replacement is a simple matter of sliding out the diffusing panels and inserting new tubes in the lamp holders. (Use deluxe warm-white or standard warm-white fluorescent tubes for light most nearly approximating the color of incandescent light.)

If insects and large dust particles settle on the diffusers, you can usually just shake them off the panels. But for thorough cleaning, take the panels out and wash them with a mild, warm detergent solution; then rinse with warm water and allow them to drain and dry naturally before replacing in the metal framework. Don't wipe them.

Panels that become warped must be replaced (though this is rarely necessary). If by some fluke one of them is cracked, it can be mended—but not invisibly—by coating the edges with plastic mending adhesive and pressing together for several hours. Don't try to mend cracked glass, however.

INDEX